ENCYCLOPEDIA OF MATHEMATICS AND ITS APPLICATIONS

EDITED BY G.-C. ROTA

Editorial Board
R. Doran, M. Ismail, T.-Y. Lam, E. Lutwak

Volume 56

Probabilistic Methods in Combinatorial Analysis

ENCYCLOPEDIA OF MATHEMATICS AND ITS APPLICATIONS

ENCYCLOPEDIA OF MATHEMATICS AND ITS APPLICATIONS

Probabilistic methods in combinatorial analysis

Vladimir N. Sachkov

*Academician of the International
Information Academy*

CAMBRIDGE
UNIVERSITY PRESS

CAMBRIDGE UNIVERSITY PRESS
Cambridge, New York, Melbourne, Madrid, Cape Town, Singapore,
São Paulo, Delhi, Dubai, Tokyo, Mexico City

Cambridge University Press
The Edinburgh Building, Cambridge CB2 8RU, UK

Published in the United States of America by Cambridge University Press, New York

www.cambridge.org
Information on this title: www.cambridge.org/9780521172776

First published 1997
First paperback edition 2010

A catalogue record for this publication is available from the British Library

Library of Congress Cataloguing in Publication data

Sachkov, Vladimir Nikolaevich.
Probabilistic methods in discrete mathematics / Vladimir N. Sachkov.
p.   cm. – (Encylopedia of mathematics and its applications; 56)
Includes bibliographical references and index.
ISBN 0 521 45512 X
1. Probabilities.   1. Title.   11. Series: Encylopeadia of mathematics
and its applications; v. 56.
QA273.S23   1995
519.2–dc20   94-30891 CIP

ISBN 978-0-521-45512-1 Hardback
ISBN 978-0-521-17277-6 Paperback

For Nevenka, Carolyn and Vesna

Contents

Preface

During the last two decades a vast number of articles have been published in which probabilistic methods have been used successfully to solve combinatorial problems, and especially to obtain various asymptotic results concerning some or other characteristic of combinatorial objects. This book is aimed at readers interested in problems of this kind both from the theoretical point of view and from the point of view of possible applications.

The book may be used by students and postgraduates in combinatorics and other fields where asymptotic methods of probability theory are applied. In particular, the material contained in this book could be taught in courses on combinatorial structures such as graphs, trees and mappings with an emphasis on the asymptotic properties of their characterictics. We believe that the asymptotic results presented here provide specialists in probability theory with new examples of applications of general limit theorems.

This text assumes a standard graduate-level knowledge of probability theory and an aquaintance with typical facts drawn from a general introduction to functions of complex variables. For the reader's convenience, the relevant results of probability theory are briefly reviewed in Chapter 1. The preliminaries from combinatorial analysis are given in the introductions to each of the subsequent chapters. Readers who are interested in obtaining more detailed knowledge of the corresponding aspects of combinatorics are advised to study my book *'Combinatorial Methods in Discrete Mathematics'* or any other basic course devoted to this subject.

The individual chapters are independent to a certain extent and the reader interested in some particular problems will have no difficulty in their study if he or she turns directly to the corresponding sections. A list of the references mentioned in the text is given at the end of the book and may be used for the study of those combinatorial problems which could not be covered within the framework of the present text.

V.N. Sachkov

Preface to the English translation

The English translation differs slightly from the Russian edition of the book in the following. I have rewritten the Introduction, Subsections 2.2, 2.3 and 6.2.2. A number of minor changes have been made throughout the text to eliminate misprints and awkward proofs. Also, the list of references has been extended by the inclusion of articles and monographs devoted to relevant problems of probabilistic combinatorics which have appeared subsequent to publication of the Russian version of the book.

I am greatly indebted to Professor B. Bollobás and Professor V. Vatutin for their help and valuable advice during the preparation of the manuscript.

V.N. Sachkov

Introduction

Many branches of mathematics owe a debt to classical combinatorics. This is especially true of probability theory. Plenty of good examples show how combinatorial considerations lead to very deep and difficult probabilistic results. The links between combinatorial and probabilistic problems have played an important role in forming probability theory as a mathematical discipline, and now manifest themselves in elementary courses devoted to this subject. The initial stage of the development of probability theory was characterized by the essential contribution of combinatorial methods in forming the mathematical background of the science. The current situation is quite different: well-developed probabilistic methods find a wide range of applications in solving various combinatorial problems. This is revealed in the search for asymptotic results in combinatorial analysis, where the probabilistic formulations of combinatorial problems provide the possibility to use the working system of notions of probability theory effectively and to take advantage of the powerful techniques of limit theorems in finding asymptotic formulae. It is appropriate to mention here that asymptotic results play an essential role in combinatorial analysis: they simplify calculations in problems oriented to applications and present the whole picture of investigated phenomena in a more transparent form.

For convenience of references some basic notions and facts of probability theory are listed in the first chapter of the book. Although these facts are presented in a systematic and unified form, this part of the monograph is not assumed to be a substitute for a textbook on probability theory, but is directed to those readers who have some basic knowledge of the subject. Preference is given to those notions and theorems which will be used in subsequent chapters and thus collecting the material, we have paid particular attention to the discrete probability theory which has the closest links with combinatorial analysis. The first chapter also includes a number of results concerning continuous distributions which are used later in obtaining some asymptotic relations.

The main objects of investigation in subsequent chapters are nonnegative matrices, partitions of finite sets, mappings of finite sets and, in particular, permutations and graphs. We also consider equivalence classes specified on the sequences of a given length constituted by elements of partially ordered sets. These combinatorial structures include the so-called allocations of objects into cells, with various restrictions on the form of possible allocations that describe whether the objects and (or) cells are distinguishable or not. Each time, a probability distribution (as a rule, a uniform one) is specified on the set of objects under consideration. The main problems in studying these objects are to find interrelations between the exact and asymptotic distributions of random variables associated with these objects. In essence, this is a way of solving combinatorial problems, and the well-established methods of combinatorial analysis are much applicable in finding the exact distribution. Probabilistic methods appear to be useful just at the stage of searching for asymptotic distributions.

Evidently, V.L. Goncharov was the first Russian mathematicians to become interested in problems of probabilistic combinatorics. In 1944 he published his paper 'Some facts from combinatorics', in which he used the method of generating functions to prove a number of limit theorems concerning series in random sequences and cycles in random permutations.

After this publication a large number of papers have appeared which study probability problems related to nonnegative matrices, partitions of sets, permutations and so on. The surveys [65, 104, 114] give a complete description of the contribution of the Russian mathematical school to the field of such problems. However, only a few monographs have been devoted to the subject. We hope that this book will fill the gap to some extent.

It should be noted that some of the results discussed in this book are given in simplified form. The reason of such simplifications is to facilitate readers the understanding of the essence of the used methods and to save them a possibility of independent estimation the permissibility of these or those generalizations. In difficult cases readers are advised to consult the literature cited at the end of the book.

CHAPTER 1

Relevant Elements of Probability Theory

1.0 Introduction

The aim of the first two sections of this chapter is to provide a survey of the
basic notions and results of probability theory which can be found in many
textbooks. The concepts and theorems mentioned in Sections 1 and 2 are
of an auxiliary nature and are included more for reference than for primary
study. For this reason the majority of statements are given without proofs,
the single exception being the theorem by Curtiss [21], which will be used
frequently throughout the book.

Section 3 deals with typical examples of applications of various limit theorems to the analysis of asymptotic distributions in combinatorial problems.
In terms of the properties of double generating functions we formulate rather
general conditions providing asymptotic normality of certain classes of probability distributions which are met in combinatorial probability. Section 4
contains a comprehensive description of limiting distributions of random
variables specified by double generating functions of the form $\exp\{x\,g(t)\}$,
where $g(t)$ is a polynomial. In subsequent chapters, the method used to
obtain the description will be extended to double generating functions of
the form $\exp\{g(x,t)\}$, where the function $g(x,t)$ is not necessarily a polynomial in t.

1.1 Probability distributions and random variables

1.1.1 Probability space

A *space* of elementary events is an arbitrary nonempty set Ω representing
all possible outcomes of an experiment which may be repeated, at least
in principle, infinitely many times. Elements of Ω are called *elementary
events*.

An *event* is any subset $A \subseteq \Omega$. An event A is said to be a particular case of an event B if the inclusion $A \subseteq B$ occurs. The *sum* $A \cup B$ of two events A and B is the event consisting of all elementary events belonging to at least one of the events A or B. The *product* $A \cap B$ is defined as the event consisting of all elementary events belonging to both A and B. The *difference* $A \smallsetminus B$ is the event consisting of the elements of the set A not belonging to B. The event $\bar{A} = \Omega \smallsetminus A$ is called the *complement* of A (that is, the *opposite* of the event A). We call the event Ω *certain*; the complement to Ω is the empty set $\varnothing = \bar{\Omega}$ and is called the *impossible* event. Events A and B are *mutually exclusive* or *disjoint* if $A \cap B = \varnothing$. Events $B_1, B_2, \ldots, B_n, \ldots$ constitute a *complete set of events* if

$$B_1 \cup B_2 \cup \ldots \cup B_n \cup \ldots = \Omega, \qquad B_i \cap B_j = \varnothing, \quad i \neq j.$$

A space of elementary events is called *discrete* if it is finite or countably infinite: $\Omega = \{\omega_1, \omega_2, \ldots, \omega_N\}$ or $\Omega = \{\omega_1, \omega_2, \ldots, \omega_n, \ldots\}$. On a discrete space Ω, let a nonnegative function $\mathbf{P}(\omega)$, $\omega \in \Omega$, be given such that $\sum_{\omega \in \Omega} \mathbf{P}(\omega) = 1$. The *probability* $\mathbf{P}(A)$ of an event $A \subseteq \Omega$ is determined by the formula

$$\mathbf{P}(A) = \sum_{\omega \in A} \mathbf{P}(\omega).$$

It follows directly from the definition of $\mathbf{P}(A)$ that

$$\mathbf{P}(\Omega) = 1, \qquad \mathbf{P}(\varnothing) = 0,$$

and, for any event A,

$$0 \leq \mathbf{P}(A) \leq 1, \qquad \mathbf{P}(\bar{A}) = 1 - \mathbf{P}(A).$$

Throughout, we make use of the following notations:

$$\bigcup_{i=1}^{n} A_i = A_1 \cup A_2 \cup \ldots \cup A_n, \qquad \bigcap_{i=1}^{n} A_i = A_1 \cap A_2 \cap \ldots \cap A_n.$$

For any finite number of events $A_1, A_2, \ldots, A_n$, *Boole's inequality*

$$\mathbf{P}\left(\bigcup_{i=1}^{n} A_i \right) \leq \sum_{i=1}^{n} \mathbf{P}(A_i)$$

is valid. If $A_1, A_2, \ldots, A_n$ are pairwise disjoint events, we have

$$\mathbf{P}\left(\bigcup_{i=1}^{n} A_i \right) = \sum_{i=1}^{n} \mathbf{P}(A_i).$$

For arbitrary events $A_1, A_2, \ldots, A_n$ the formula

(1.1) $$\mathbf{P}\left(\bigcup_{i=1}^{n} A_i\right) = \sum_{k=1}^{n} (-1)^{k-1} S_k$$

is valid, where

$$S_k := \sum_{1 \le i_1 < \cdots < i_k \le n} \mathbf{P}\left(A_{i_1} A_{i_2} \ldots A_{i_k}\right),$$

and summation is accomplished over all possible combinations of k numbers taken from $1, 2, \ldots, n$ without repetition, and

$$A_{i_1} A_{i_2} \ldots A_{i_k} = \bigcap_{j=1}^{k} A_{i_j}.$$

Let Ω be a discrete space of elementary events. Any numerical function $\xi := \xi(\omega)$ of $\omega \in \Omega$ is called a *random variable*. A random variable is said to be *discrete* if the set of its values is finite or countably infinite.

The *distribution law* $\mathbf{P}_\xi(B)$ of the discrete random variable ξ is the probability

$$\mathbf{P}_\xi(B) = \mathbf{P}\{\omega : \xi(\omega) \in B\},$$

defined for each number set B. In particular, for $x \in (-\infty, \infty)$ we write

$$\mathbf{P}_\xi(x) = \mathbf{P}\{\omega : \xi(\omega) = x\} = \mathbf{P}\{\xi = x\},$$

where the notation $\{\xi = x\}$ stands for the event $\{\omega : \xi(\omega) = x\}$.

In what follows we will use mainly discrete spaces of elementary events and discrete random variables. However, in order to study limiting probability laws, we must consider arbitrary spaces of elementary events.

A system of subsets $\mathfrak{F}$ of an arbitrary space of elementary events Ω is called a *σ-algebra* if the following conditions are satisfied:

(1) $\Omega \in \mathfrak{F}$;

(2) If $A \in \mathfrak{F}$ then $\bar{A} \in \mathfrak{F}$;

(3) If $\{A_n\}$ is a sequence of sets from $\mathfrak{F}$ then

$$\bigcup_{i=1}^{\infty} A_i \in \mathfrak{F}, \qquad \bigcap_{i=1}^{\infty} A_i \in \mathfrak{F}$$

(obviously, the fulfilment of only one of the conditions is sufficient since the others will follow as a corollary).

If $\mathfrak{F}$ is a σ-algebra of Ω then the pair $(\Omega, \mathfrak{F})$ is said to be a *measurable space*. Elements of the σ-algebra $\mathfrak{F}$ are called the events corresponding to the measurable space $(\Omega, \mathfrak{F})$.

A numerical function $\mathbf{P}$ defined on the σ-algebra $\mathfrak{F}$ of a measurable space $(\Omega, \mathfrak{F})$ is called a *probability* if it satisfies the following axioms:

(1) $\mathbf{P}(A) \geq 0$ for any $A \in \mathfrak{F}$.

(2) For any sequence $\{A_n\}$ of pairwise disjoint events the following equality is valid:

$$\mathbf{P}\left(\bigcup_{i=1}^{\infty} A_i\right) = \sum_{i=1}^{\infty} \mathbf{P}(A_i).$$

(3) $\mathbf{P}(\Omega) = 1$.

The triple $(\Omega, \mathfrak{F}, \mathbf{P})$ is called a *probability space*. Any probability space satisfying the axioms mentioned possesses all the properties listed for discrete spaces of elementary events. Let $(\Omega, \mathfrak{F}, \mathbf{P})$ be a probability space and let $A, B \in \mathfrak{F}$. If $\mathbf{P}(B) > 0$ then the number $\mathbf{P}(A\,|\,B)$, defined by the formula

$$\mathbf{P}(A\,|\,B) := \frac{\mathbf{P}(A \cap B)}{\mathbf{P}(B)},$$

is called the *conditional* probability of the event A under the condition B.

If events $B_1, B_2, \ldots, B_n, \ldots$ are such that $B_i \cap B_j = \emptyset$, $\mathbf{P}(B_i) > 0$ and $A \subseteq \cup B_i$, then the following formula is valid (the *total probability* formula):

$$\mathbf{P}(A) = \sum_{i=1}^{\infty} \mathbf{P}(B_i)\,\mathbf{P}(A\,|\,B_i).$$

Events $A_1, A_2, \ldots, A_n$ are said to be *mutually independent* if, for all combinations of the subscripts $1 \leq i_1 < \cdots < i_k \leq n$, $k = 2, 3, \ldots, n$, we have

$$\mathbf{P}\left(\bigcap_{s=1}^{k} A_{i_s}\right) = \prod_{s=1}^{k} \mathbf{P}(A_{i_s}).$$

We now give the definition of a random variable for an arbitrary probability space. A *random variable* ξ on a probability space $(\Omega, \mathfrak{F}, \mathbf{P})$ is a numerical function $\xi := \xi(\omega)$, $\omega \in \Omega$, mapping Ω onto the set of real numbers $\mathbf{R}$ such that for any $x \in (-\infty, \infty)$ the set $\{\omega\colon \xi(\omega) < x\}$ belongs to the σ-algebra $\mathfrak{F}$. The function $F_\xi(x) := \mathbf{P}\{\xi < x\}$ is called the *distribution function* or, simply, the *distribution* of the random variable ξ. Any distribution function $F_\xi(x)$ is defined for all $x \in (-\infty, \infty)$; it is nondecreasing and left-continuous. Also,

$$\lim_{x \to -\infty} F_\xi(x) = 0, \qquad \lim_{x \to \infty} F_\xi(x) = 1.$$

If ξ is a discrete random variable taking values $x_1 < x_2 < \cdots$ with positive probabilities, then

$$\mathbf{P}\{\xi = x_k\} = F_\xi(x_{k+1}) - F_\xi(x_k), \qquad k = 1, 2, \ldots.$$

A random variable ξ is said to be *continuous* if there exists a function $p_\xi(x) \geq 0$ such that, for all $x \in (-\infty, \infty)$,

$$F_\xi(x) = \int_{-\infty}^{x} p_\xi(y)\, dy, \qquad \int_{-\infty}^{\infty} p_\xi(y)\, dy = 1.$$

The function $p_\xi(x)$ is called the *density* of the distribution of the random variable ξ.

A random variable ξ is said to be normally distributed with parameters (m, σ) if its distribution function $\Phi_{m,\sigma}(x)$ has the form

$$(1.2) \qquad \Phi_{m,\sigma}(x) = \frac{1}{\sqrt{2\pi}\,\sigma} \int_{-\infty}^{x} \exp\left\{ -\frac{(u-m)^2}{2\sigma^2} \right\} du$$

with $\sigma > 0$. If $m = 0$ and $\sigma = 1$ we say that ξ has *standard normal distribution*.

Let $\xi_1, \xi_2, \ldots, \xi_n$ be random variables defined on a common probability space $(\Omega, \mathfrak{F}, \mathbf{P})$. The vector $(\xi_1, \xi_2, \ldots, \xi_n)$ is called an *n-dimensional random variable* or a *random vector*. The function

$$F_{\xi_1 \ldots \xi_n}(x_1, \ldots, x_n) = \mathbf{P}\left\{ \xi_1 < x_1, \ldots, \xi_n < x_n \right\},$$

$$x_i \in (-\infty, \infty), \qquad i = 1, 2, \ldots, n,$$

is the distribution function of the *n*-dimensional random variable $(\xi_1, \xi_2, \ldots, \xi_n)$. A random variable (vector) $(\xi_1, \xi_2, \ldots, \xi_n)$ is said to be *continuous* if there exists a function $p_{\xi_1 \ldots \xi_n}(y_1, \ldots, y_n) \geq 0$, $y_i \in (-\infty, \infty)$, $i = 1, 2, \ldots, n$, such that

$$\int_{-\infty}^{\infty} \cdots \int_{-\infty}^{\infty} p_{\xi_1 \ldots \xi_n}(y_1, \ldots, y_n)\, dy_1 \ldots dy_n = 1$$

and

$$F_{\xi_1 \ldots \xi_n}(x_1, \ldots, x_n) = \int_{-\infty}^{x_1} \cdots \int_{-\infty}^{x_n} p_{\xi_1 \ldots \xi_n}(y_1, \ldots, y_n)\, dy_1 \ldots dy_n.$$

The function $p_{\xi_1 \ldots \xi_n}(y_1, \ldots, y_n)$ is called the *density* of the distribution of the *n*-dimensional random variable $(\xi_1, \xi_2, \ldots, \xi_n)$ or, simply, the density of $(\xi_1, \xi_2, \ldots, \xi_n)$.

The distribution of a continuous random variable $(\xi_1, \xi_2, \ldots, \xi_n)$ is said to be a *nondegenerate normal distribution* if its density has the form

$$(1.3) \qquad p_{\xi_1 \ldots \xi_n}(x_1, \ldots, x_n) = (\det A)^{1/2} \, (2\pi)^{-n/2}$$

$$\times \exp\left\{-\tfrac{1}{2} Q(x_1, \ldots, x_n)\right\},$$

where

$$Q(x_1, \ldots, x_n) := \sum_{i,j=1}^{n} a_{ij} \, x_i \, x_j$$

is a positive definite quadratic form and $\det A$ is the determinant of the matrix $A = \| a_{ij} \|$.

The random variables $\xi_1, \xi_2, \ldots, \xi_n$ are said to be *independent* if

$$F_{\xi_1 \ldots \xi_n}(x_1, \ldots, x_n) = F_{\xi_1}(x_1) \, \cdots \, F_{\xi_n}(x_n)$$

for all tuples $(x_1, \ldots, x_n)$, $x_i \in (-\infty, \infty)$, $i = 1, \ldots, n$.

The random variables of an infinite sequence $\{\xi_k\}$ are said to be independent if the preceding equality holds for any n. If an n-dimensional random vector $(\xi_1, \xi_2, \ldots, \xi_n)$ is continuous and the random variables $\xi_1, \xi_2, \ldots, \xi_n$ are independent then the densities of the variables are related by the formula

$$p_{\xi_1 \ldots \xi_n}(x_1, \ldots, x_n) = p_{\xi_1}(x_1) \, \cdots \, p_{\xi_n}(x_n),$$

$$x_i \in (-\infty, \infty), \quad i = 1, 2, \ldots, n.$$

For discrete independent random variables $\xi_1, \xi_2, \ldots, \xi_n$ the equality

$$\mathbf{P}\{\xi_1 = x_1, \ldots, \xi_n = x_n\} = \mathbf{P}\{\xi_1 = x_1\} \, \cdots \, \mathbf{P}\{\xi_n = x_n\}$$

is valid for all $x_i \in (-\infty, \infty)$, $i = 1, \ldots, n$.

1.1.2 Moments of random variables

First we recall the definition of the Stieltjes integral for a distribution function $F(x)$ and a function $f(x)$ continuous on an interval $[a, b]$ of the real line. We partition the interval $[a, b]$ into n subintervals $[x_i, x_{i+1}]$ such that $a = x_0 < x_1 < x_2 < \cdots < x_n = b$ and calculate a sum

$$S := \sum_{i=1}^{n} f(\tilde{x}_i) \left[F(x_i) - F(x_{i-1}) \right],$$

where $\tilde{x}_i$ is any number from the interval $[x_i,\,x_{i+1}]$. If the sum S tends to a finite limit as

$$\max_{1\le i\le n}\,|\,x_{i+1}-x_i\,|\longrightarrow 0,\qquad\qquad n\to\infty,$$

and this limit does not depend on the particular sequence of partitions and on the choice of the points $\tilde{x}_i$, then it is called the Stieltjes integral of $f(x)$ with respect to the distribution function $F(x)$ and is denoted by

$$\int_a^b f(x)\,dF(x).$$

In what follows we assume that the Stieltjes integral of a function $f(x)$ with respect to a function $F(x)$ exists if and only if the corresponding integral of the function $|f(x)|$ exists. By definition,

$$\int_{-\infty}^{\infty} f(x)\,dF(x)=\lim_{\substack{a\to-\infty\\ b\to\infty}}\int_a^b f(x)\,dF(x).$$

Let ξ be a random variable defined on a probability space $\left(\Omega,\,\mathfrak{F},\,\mathbf{P}\right)$. The *mathematical expectation* (or mean value or, simply, mean) of the random variable ξ is the number

$$\mathbf{E}\,\xi := \int_{-\infty}^{\infty} x\,dF(x),$$

where $F(x)$ is the distribution function of ξ. The mathematical expectation of a discrete random variable ξ taking values $\cdots < x_{-1} < x_0 < x_1 < \cdots < x_n < \cdots$ is calculated by the formula

$$\mathbf{E}\,\xi := \sum_{k=-\infty}^{\infty} x_k\,\mathbf{P}\,\{\xi = x_k\};$$

if ξ is a continuous random variable with density $p_\xi(x)$, the integral in the definition of expectation reduces to the usual Riemann integral, namely,

$$\mathbf{E}\,\xi := \int_{-\infty}^{\infty} x\,p_\xi(x)\,dx.$$

We list the basic properties of expectation:

(1) $\mathbf{E}\,(C\,\xi) = C\,\mathbf{E}\,\xi$ for any constant C;

(2) $\mathbf{E}\,(\xi_1+\xi_2) = \mathbf{E}\,\xi_1 + \mathbf{E}\,\xi_2$ if the mathematical expectations $\mathbf{E}\,\xi_1$ and $\mathbf{E}\,\xi_2$ exist;

(3) If ξ_1 and ξ_2 are independent random variables then

$$\mathbf{E}\,\xi_1\xi_2 = \mathbf{E}\,\xi_1 \cdot \mathbf{E}\,\xi_2.$$

The *variance* of a random variable ξ is defined by the formula

$$\operatorname{Var}\xi := \mathbf{E}\left(\xi - \mathbf{E}\,\xi\right)^2.$$

Variance has the following basic properties:

(1) $\operatorname{Var}(C\,\xi) = C^2\operatorname{Var}\xi$ for any constant C;

(2) If ξ_1 and ξ_2 are independent random variables then

$$\operatorname{Var}(\xi_1 + \xi_2) = \operatorname{Var}\xi_1 + \operatorname{Var}\xi_2;$$

(3) Let ξ be a nonnegative random variable: $\xi \geq 0$, and let ε be an arbitrary positive number. The following inequality is valid:

$$(1.4) \qquad \mathbf{P}\{\xi \geq \varepsilon\} \leq \frac{\mathbf{E}\,\xi}{\varepsilon}.$$

This inequality implies, for any random variable ξ, the *Chebyshev inequality*:

$$(1.5) \qquad \mathbf{P}\{|\xi - \mathbf{E}\,\xi| \geq \varepsilon\} \leq \frac{\operatorname{Var}\xi}{\varepsilon^2}.$$

The *moment of the kth order* of a random variable ξ is defined to be the quantity $M_k := \mathbf{E}\,\xi^k$ (if the mathematical expectation exists). The numbers $\mathbf{E}\,|\xi|^k$ and $\mu_k := \mathbf{E}(\xi - \mathbf{E}\,\xi)^k$ are called, respectively, the *absolute* and *central moments* of order k. The following relations are valid:

$$\mu_k = \sum_{j=0}^{k} (-1)^{k-j}\binom{k}{j} M_j\, M_1^{k-j}, \qquad k = 0, 1, \ldots,$$

$$M_k = \sum_{j=0}^{k} \binom{k}{j}\mu_j\, M_1^{k-j}, \qquad k = 0, 1, \ldots,$$

where $\mu_0 = M_0 = 1$. The *factorial* and *binomial moments* of order k are defined, respectively, by the equalities

$$[M]_k := \mathbf{E}\,(\xi)_k, \qquad k = 0, 1, \ldots,$$

$$B_k := \mathbf{E}\binom{\xi}{k}, \qquad k = 0, 1, \ldots,$$

where

$$(\xi)_0 := 1, \quad (\xi)_k := \xi(\xi-1)\cdots(\xi-k+1), \quad k > 0, \quad \binom{\xi}{k} := \frac{(\xi)_k}{k!}.$$

Obviously,

$$[M]_k = k!\, B_k, \qquad\qquad k = 0, 1, \ldots.$$

The following relations are valid:

$$[M]_k = \sum_{j=0}^{k} s(k,j)\, M_j, \qquad\qquad k = 0, 1, \ldots,$$

$$M_k = \sum_{j=0}^{k} \sigma(k,j)\, [M]_j, \qquad\qquad k = 0, 1, \ldots,$$

where $s(k,j)$ and $\sigma(k,j)$ are the *Stirling numbers* of the *first* and *second kind* respectively. These numbers are defined by the equalities

$$(x)_k := \sum_{j=0}^{k} s(k,j)\, x^j, \qquad\qquad k = 0, 1, \ldots,$$

$$x^k := \sum_{j=0}^{k} \sigma(k,j)\, (x)_j, \qquad\qquad k = 0, 1, \ldots.$$

1.1.3   Integer-valued random variables

Discrete random variables taking only integer values will be of special importance in this book. Such random variables are called *integer-valued*. Below we consider the case of nonnegative integer-valued random variables.

Let ξ be a nonnegative integer-valued random variable with

$$q_k := \sum_{j=k+1}^{\infty} \mathbf{P}\{\xi = j\}, \qquad\qquad k = 0, 1, \ldots.$$

Then we have

(1.6) $$\mathbf{E}\,\xi = \sum_{k=0}^{\infty} q_k.$$

The *generating function* of an integer-valued random variable ξ is defined by the equality

$$P(x) := \sum_{k=0}^{\infty} P_k\, x^k, \qquad P_k := \mathbf{P}\{\xi = k\}.$$

It is clear that $P(x)$ is an analytic function within the circle $|x| \leq 1$ and, in view of the equality

$$P_k = \frac{1}{k!}\, P^{(k)}(0), \qquad k = 0, 1, \ldots,$$

where $P^{(k)}(0)$ is the value of the kth derivative of $P(x)$ at the point $x = 0$, it determines the distribution of ξ uniquely. One can use, as an inversion formula, *Cauchy's integral formula*:

$$P_k = \frac{1}{2\pi i} \oint_{C} P(z)\, \frac{dz}{z^{n+1}},$$

where C is a contour in the complex plane enclosing the origin and lying inside the circle where $P(x)$ is analytic.

Let $\xi_1, \xi_2, \ldots, \xi_n$ be independent random variables and let $P_1(x)$, $P_2(x), \ldots, P_n(x)$ be the corresponding generating functions. The generating function $P(x)$ of the random variable $\zeta = \xi_1 + \xi_2 + \cdots + \xi_n$ is given by the formula

$$P(x) = P_1(x)\, P_2(x)\, \cdots\, P_n(x).$$

For a random variable ξ the functions

$$M(x) := \sum_{k=0}^{\infty} M_k\, \frac{x^k}{k!},$$

$$\bar{M}(x) := \sum_{k=0}^{\infty} [M]_k\, \frac{x^k}{k!},$$

$$B(x) := \sum_{k=0}^{\infty} B_k\, x^k$$

are called, respectively, the *moment generating function*, the *factorial moment generating function* and the *binomial moment generating function* of ξ. These generating functions are expressed by the generating function of the random variable ξ as follows:

(1.7) $$M(x) = P(e^x),$$

(1.8) $$\bar{M}(x) = B(x) = P(x + 1).$$

Binomial moments of ξ can be computed by the values of derivatives of $P(x)$ at the point $x = 1$:

$$B_k = \frac{1}{k!} P^{(k)}(1), \qquad k = 0, 1, \dots .$$

Let us consider some examples of integer-valued variables.

1.1.3.1 Binomial distribution

The random variable

$$\zeta^{(n)} = \xi_1 + \xi_2 + \cdots + \xi_n ,$$

where $\xi_1, \xi_2, \dots, \xi_n$ are independent and $\mathbf{P}\{\xi_i = 1\} = p$, $\mathbf{P}\{\xi_i = 0\} = q$, $p + q = 1$, is said to have *binomial distribution* with parameters (n, p). Obviously,

$$\mathbf{P}\{\zeta^{(n)} = k\} = \binom{n}{k} p^k q^{n-k}, \qquad k = 0, 1, \dots, n .$$

The generating function and the binomial moment generating function of $\zeta^{(n)}$ are of the form:

(1.9) $$P_n(x) = (px + q)^n,$$

(1.10) $$B_n(x) = (px + 1)^n.$$

Hence the formula

$$B_k = \binom{n}{k} p^k, \qquad k = 0, 1, \dots, n ,$$

for binomial moments of $\zeta^{(n)}$ follows.

1.1.3.2 Pascal distribution

Consider a scheme of independent Bernoulli trials. Each elementary trial has probability p of success and q of failure, $p + q = 1$. Let ξ be the number of failures till the rth success occurs. The law of distribution of the random variable ξ,

(1.11) $$\mathbf{P}\{\xi = k\} = \binom{k + r - 1}{k} p^r q^k, \qquad k = 0, 1, \dots ,$$

is called a *Pascal distribution*. The generating function and the binomial moment generating function of ξ are

$$(1.12) \qquad P(x) = \left(\frac{p}{1 - q\,x} \right)^r ,$$

$$(1.13) \qquad B(x) = \left(\frac{p}{p - q\,x} \right)^r .$$

Binomial moments can be calculated by the formula

$$(1.14) \qquad B_k = \binom{r + k - 1}{k} \left(\frac{q}{p} \right)^k , \qquad k = 0, 1, \dots .$$

1.1.3.3 Poisson distribution

A random variable ξ is said to have *Poisson distribution* with parameter $\lambda > 0$ if

$$\mathbf{P}\{\xi = k\} = \frac{\lambda^k}{k!}\, e^{-\lambda}, \qquad k = 0, 1, \dots .$$

Using the generating function of the distribution of ξ,

$$(1.15) \qquad P(x) = e^{\lambda (x - 1)},$$

and the binomial moment generating function of ξ,

$$(1.16) \qquad B(x) = e^{\lambda x},$$

we find the binomial moments of ξ:

$$(1.17) \qquad B_k = \frac{\lambda^k}{k!}, \qquad k = 0, 1, \dots .$$

1.1.3.4 Hypergeometric distribution

Consider a set consisting of n elements, m of which are of one kind and $n - m$ of the other. Denote by ξ the number of elements of the first kind contaned in a sample of size r chosen at random from the set. The distribution of the random variable ξ,

$$\mathbf{P}\{\xi = k\} = \binom{m}{k} \binom{n - m}{r - k} \Big/ \binom{n}{r} , \qquad k = 0, 1, \dots, \min(m, r),$$

is called *hypergeometric*. The binomial moments of the distribution have the form

$$(1.18) \qquad B_k = \binom{m}{k} \binom{n - k}{r - k} \Big/ \binom{n}{r} , \qquad k = 0, 1, \dots, \min(m, r).$$

1.1.4   Inversion formulae

Let ξ be a random variable taking integer values not exceeding n. Then relation (1.8) yields

$$P(x) = \sum_{k=0}^{n} B_k \, (x-1)^k = \sum_{r=0}^{n} x^r \sum_{k=r}^{n} (-1)^{k-r} \binom{k}{r} B_k \, .$$

Hence we deduce the following formula expressing the probability distribution of the random variable ξ by its binomial moments:

$$(1.19) \qquad \mathbf{P}\{\xi = r\} = \sum_{k=r}^{n} (-1)^{k-r} \binom{k}{r} B_k \, , \qquad\qquad r = 0, 1, \ldots, n \, .$$

This formula may be viewed as the inverse of the formula

$$B_k = \sum_{j=k}^{n} \binom{j}{k} \mathbf{P}\{\xi = j\} \, , \qquad\qquad k = 0, 1, \ldots, n \, ,$$

which allows to recover the binomial moments of ξ from its distribution.

Let Ω be a finite space of elementary events and P_r be the probability of the event that in a random trial exactly r events occur among the events $A_1, A_2, \ldots, A_n$. The following formulae are valid (they are usually called the formulae of the *inclusion–exclusion principle*):

$$(1.20) \qquad \begin{cases} P_r &= \sum_{k=r}^{n} (-1)^{k-r} \binom{k}{r} S_k \, , \qquad r = 0, 1, \ldots, \\[2mm] S_0 &= 1 \, , \qquad S_k = \sum_{1 \le j_1 < \cdots < j_k \le n} \mathbf{P}\big(A_{j_1} \ldots A_{j_k} \big) \, . \end{cases}$$

Indeed, let ξ_i be a random variable which is equal to 1 if the event A_i occurs, and is equal to 0 otherwise. If $\xi = \xi_1 + \xi_2 + \cdots + \xi_n$ then $P_r = \mathbf{P}\{\xi = r\}$. Observe that

$$(1.21) \qquad \binom{\xi}{k} = \sum_{k_1 + \cdots + k_n = k} \binom{\xi_1}{k_1} \binom{\xi_2}{k_2} \cdots \binom{\xi_n}{k_n} \, ,$$

where the sum is taken over all tuples $(k_1, k_2, \ldots, k_n)$ with integer-valued nonnegative elements satisfying the condition $k_1 + k_2 + \cdots + k_n = k$. Taking the mathematical expectation of both sides of equality (1.21) we find the kth binomial moment of ξ:

$$(1.22) \qquad B_k = \sum_{1 \le j_1 < \cdots < j_k \le n} \mathbf{P}\{\xi_{j_1} = 1, \ldots, \xi_{j_k} = 1\} \, .$$

Now formulae (1.20) follow from (1.19) and (1.22).

1.1.4.1 Matching problem or 'probléme des rencontres'

Let $(a_1, a_2, \ldots, a_n)$ be a permutation chosen at random with probability $1/n!$ from the set of all possible permutations of numbers $1, 2, \ldots, n$. The problem is to find the distribution of the number ξ of 'rencontre', that is, the number of j, $1 \leq j \leq n$, such that $a_j = j$. We consider the events $A_1, A_2, \ldots, A_n$, where the event A_j means that $a_j = j$. Clearly,

$$\mathbf{P}\left(A_{j_1} A_{j_2} \cdots A_{j_k} \right) = \frac{(n-k)!}{n!}, \qquad 1 \leq j_1 < \cdots < j_k \leq n.$$

From (1.20) we find that

$$P_r = \mathbf{P}\{\xi = r\} = \frac{1}{r!} \sum_{k=0}^{n-r} (-1)^k \frac{1}{k!}, \qquad r = 0, 1, \ldots, n.$$

Formulae (1.20) imply Bonferroni's inequalities

$$\sum_{k=r}^{r+2\nu-1} (-1)^{k-r} \binom{k}{r} S_k \leq P_r \leq \sum_{k=r}^{r+\nu} (-1)^{k-r} \binom{k}{r} S_k,$$

where $\nu \geq 1$ is an arbitrary integer satisfying the condition $r + 2\nu \leq n$. We are able to refine the sense of Bonferroni's inequalities a little. Retaining only the terms containing $S_r, S_{r+1}, \ldots, S_{r+t}$ in (1.20) and removing those containing $S_{r+t+1}, S_{r+t+2}, \ldots, S_n$, we see that the error of the approximation does not exceed in the absolute value the first of the terms removed and its sign coincides with that of the first term.

Inversion formulae similar to (1.19) may be proved in the general case. We formulate the corresponding theorem due to Takács [116].

Theorem 1.1 *If all binomial moments B_k of a random variable ξ are finite and*

$$\rho := \limsup_{k \to \infty} B_k^{1/k} < \infty$$

then, for $r = 0, 1, \ldots$,

$$(1.23) \quad \mathbf{P}\{\xi = r\} = \sum_{k=r}^{\infty} \binom{k}{r} (1+d)^{-k-1} \sum_{j=r}^{k} (-1)^{j-r} \binom{k-r}{j-r} d^{k-j} B_j,$$

where $d > \rho^2 - 1$ is any nonnegative number. If $\rho < 1$ then one may choose $d = 0$, which reduces (1.23) to the form

$$(1.24) \qquad \mathbf{P}\{\xi = r\} = \sum_{k=r}^{\infty} (-1)^{k-r} \binom{k}{r} B_k, \qquad r = 0, 1, \ldots.$$

In fact, Theorem 1.1 describes the conditions under which a probability distribution is uniquely determined by its binomial moments and provides a formula which allows us to find the distribution by these moments.

Let us consider some examples showing how Theorem 1.1 can be used.

Example 1   For a random variable ξ having Poisson distribution with parameter λ, the binomial moment of the kth order is

$$B_k = \frac{\lambda^k}{k!}$$

and, therefore,

$$\rho = \lim_{k \to \infty} B_k^{1/k} = 0.$$

Consequently, this distribution is uniquely determined by its binomial moments and, according to formula (1.24), we have

$$\mathbf{P}\{\xi = r\} = \sum_{k=r}^{\infty} (-1)^{k-r} \binom{k}{r} \frac{\lambda^k}{k!} = \frac{\lambda^r}{r!} e^{-\lambda}, \qquad r = 0, 1, \ldots .$$

Example 2   From (1.14) it follows that, for a random variable ξ having Pascal distribution,

$$\rho = \lim_{k \to \infty} B_k^{1/k} = q/p.$$

Thus, we can apply formula (1.23) for $p > 0$. From (1.23), taking relation (1.14) into account, we deduce formula (1.11). For $q < p$ the same result follows from (1.24).

Example 3   We consider a random variable ξ whose binomial moments are of the form

$$B_k = \frac{\lambda^k}{(k!)^2}, \qquad k = 0, 1, \ldots ,$$

where $\lambda > 0$. In this case,

$$\rho = \lim_{k \to \infty} B_k^{1/k} = 0$$

and, using formula (1.24), we find

$$\mathbf{P}\{\xi = r\} = \frac{\lambda^r}{r!} \sum_{j=0}^{\infty} (-1)^j \frac{\lambda^j}{j!\,(j+r)!}, \qquad r = 0, 1, \ldots .$$

For any positive integer k, the *Bessel function* J_k is defined by the series

$$J_k(2\sqrt{z}) = z^{k/2} \sum_{j=0}^{\infty} (-1)^j \frac{z^j}{j!\,(k+j)!} .$$

Using this representation we obtain

$$\mathbf{P}\{\xi = r\} = \frac{\lambda^{r/2}}{r!} J_r(2\sqrt{\lambda}), \qquad r = 0, 1, \ldots .$$

1.1.5 Fréchet formulae

Let Ω be an arbitrary space of elementary events and let $A_1, A_2, \ldots, A_n$ be a sequence of events which may, generally speaking, be intersecting and dependent, and let P_r be the probability that exactly r events occur in a random trial. To calculate P_r one can use the following formulae originally due to Fréchet and coinciding in the case of finite Ω with formulae (1.20) of the inclusion–exclusion principle:

$$
(1.25) \quad
\begin{cases}
P_r = \sum_{k=r}^{n} (-1)^{k-r} \binom{k}{r} S_k, & r = 0, 1, \ldots, n, \\
S_0 = 1, \; S_k = \sum_{1 \le j_1 < \cdots < j_k \le n} \mathbf{P}(A_{j_1} \ldots A_{j_k}), & k = 1, 2, \ldots, n.
\end{cases}
$$

Let $(j_1, j_2, \ldots, j_n)$ be a permutation of numbers $1, 2, \ldots, n$ and let

$$
\mathfrak{U}_{j_1 \cdots j_r} := A_{j_1} \ldots A_{j_r} \bar{A}_{j_{r+1}} \ldots \bar{A}_{j_n}.
$$

The $\binom{n}{r}$ events introduced are disjoint and

$$
(1.26) \qquad P_r = \sum_{I_r} \mathbf{P}(\mathfrak{U}_{j_1 \cdots j_r}),
$$

where $I_r := \{ \langle j_1, \ldots, j_r \rangle : 1 \le j_1 < \cdots < j_r \le n \}$ is the aggregate of combinations of size r from the elements $1, 2, \ldots, n$. On the other hand,

$$
\mathbf{P}(A_{i_1} \ldots A_{i_k}) = \mathbf{P}(A_{i_1} \ldots A_{i_k} (A_{i_{k+1}} + \bar{A}_{i_{k+1}}) \ldots (A_{i_n} + \bar{A}_{i_n}))
$$

$$
= \sum_{r=k}^{n} \sum_{I_r} \mathbf{P}(\mathfrak{U}_{j_1 \cdots j_r}),
$$

where the second summation is taken over all the combinations $\langle j_1, j_2, \ldots, j_r \rangle$ containing the combination $\langle i_1, i_2, \ldots, i_k \rangle$. Since each combination $\langle i_1, i_2, \ldots, i_k \rangle$ belongs to exactly $\binom{n}{r}$ combinations of the form $\langle j_1, j_2, \ldots, j_k \rangle$, we have

$$
(1.27) \qquad S_k = \sum_{r=k}^{n} \binom{r}{k} \sum_{I_r} \mathbf{P}(\mathfrak{U}_{j_1 \cdots j_r}).
$$

Equalities (1.26) and (1.27) yield

$$
(1.28) \qquad S_k = \sum_{r=k}^{n} \binom{r}{k} P_r, \qquad r = 1, 2, \ldots, n.
$$

Multiplying both sides of (1.28) by $(-1)^{k-r}\binom{k}{r}$ and summing over $k = r, r+1, \ldots, n$, we see that formulae (1.25) are valid. We note that formula (1.1), given at the beginning of this chapter, is a corollary of Fréchet's formulae (1.25).

Let $\xi_1, \xi_2, \ldots, \xi_n$ be independent random variables, each of which is uniformly distributed on the interval $(0, 1)$. The distribution function of their sum is found by the relation

$$(1.29) \qquad \mathbf{P}\{\xi_1 + \xi_2 + \cdots + \xi_n < x\} = \frac{1}{n!}\sum_{k=0}^{[x]}(-1)^k\binom{n}{k}(x-k)^n,$$

which is known as *Laplace's formula*. We prove this fact.

Let us agree that a point $(x_1, x_2, \ldots, x_n)$ of the n-dimensional simplex

$$\Omega = \left\{(x_1, x_2, \ldots, x_n)\colon x_1 + x_2 + \cdots + x_n < x,\ x_i \geq 0,\ i = 1, 2, \ldots, n\right\}$$

possesses property A_i if $x_i > 1$. The volume of that part of Ω that possesses properties $A_{j_1}, A_{j_2}, \ldots, A_{j_k}$ simultaneously is calculated by the formula

$$(1.30) \qquad V\left(A_{j_1} \ldots A_{j_k}\right) = \begin{cases} \dfrac{(x-k)^n}{n!}, & k \leq x, \\[2mm] 0, & k > x. \end{cases}$$

Obviously, $\mathbf{P}\{\xi_1 + \xi_2 + \cdots + \xi_n < x\} = V(\Omega_0)$, where $V(\Omega_0)$ is the volume of the simplex

$$\Omega_0 = \left\{(x_1, x_2, \ldots, x_n)\colon x_1 + x_2 + \cdots + x_n < x,\ 0 \leq x_i \leq 1,\ i = 1, 2, \ldots, n\right\}.$$

On the other hand,

$$V(\Omega_0) = \sum_{k=0}^{[x]}(-1)^k S_k,$$

where $S_0 = V(\Omega)$ and

$$S_k = \sum_{1 \leq j_1 < \cdots < j_k \leq n} V\left(A_{j_1} \ldots A_{j_k}\right), \qquad k = 1, 2, \ldots.$$

Hence, taking (1.30) into account, we deduce the desired representation (1.29).

1.1.6 Continuity theorem for generating functions

Let $\xi_1, \xi_2, \ldots, \xi_n, \ldots$ be a sequence of integer-valued random variables with distribution laws $\mathbf{P}\{\xi_n = k\}$, $k = 0, 1, \ldots$; $n = 1, 2, \ldots$, and let $P_1(x), P_2(x), \ldots, P_n(x), \ldots$ be the corresponding sequence of generating functions. A probability distribution P_k, $k = 0, 1, \ldots$, is said to be *limiting* for the sequence of distributions $\mathbf{P}\{\xi_n = k\}$, $k = 0, 1, \ldots$; $n = 1, 2, \ldots$, if

$$\lim_{n \to \infty} \mathbf{P}\{\xi_n = k\} = P_k, \qquad k = 0, 1, \ldots.$$

We now formulate a theorem which is usually called the *continuity theorem* for generating functions.

Theorem 1.2 *For the equalities*

$$\lim_{n \to \infty} \mathbf{P}\{\xi_n = k\} = P_k, \qquad k = 0, 1, \ldots,$$

to be valid it is necessary and sufficient that, for any $x \in [0, 1)$,

$$\lim_{n \to \infty} P_n(x) = P(x),$$

where $P(x)$ *is the generating function of the distribution law* P_k, $k = 0, 1, \ldots$.

We consider two examples to demonstrate the importance of the theorem. Let us assume that $p = 1 - \lambda r^{-1}$, $\lambda = \text{const}$ and $r \to \infty$. Then, using formula (1.12) for the generating function of a Pascal distribution, we have

$$\lim_{r \to \infty} \left(\frac{1 - \lambda r^{-1}}{1 - \lambda\, x\, r^{-1}} \right)^r = \exp\{\lambda(x - 1)\}.$$

This means that the Poisson distribution with parameter λ is limiting for the Pascal distribution if $p = 1 - \lambda r^{-1}$ and $r \to \infty$. Similarly, if $p = \lambda n^{-1}$, $\lambda = \text{const}$ and $n \to \infty$, then, according to (1.9), the generating function of the binomial distribution with parameters $(\lambda n^{-1}, n)$ satisfies the relation

$$\lim_{n \to \infty} \left(1 + \frac{\lambda(x - 1)}{n} \right)^n = \exp\{\lambda(x - 1)\}.$$

1.2 Characteristic functions and moment generating functions

1.2.1 Properties of characteristic functions

The *characteristic function* $\varphi_\xi(t)$, $t \in (-\infty, \infty)$, of the random variable ξ having distribution function $F_\xi(x)$ is defined by the equality

$$\varphi_\xi(t) = \mathbf{E}\, e^{it\xi} = \int\limits_{-\infty}^{\infty} e^{itx}\, dF_\xi(x)\,, \qquad i = \sqrt{-1}\,,$$

where $\mathbf{E}(\xi + i\eta) = \mathbf{E}\,\xi + i\,\mathbf{E}\,\eta$ for real ξ and η and the integral is a Lebesgue–Stieltjes integral.

For example, if ξ is a random variable uniformly distributed on the segment $[0, 1]$ then its characteristic function has the form

$$\varphi_\xi(t) = \int\limits_0^1 e^{itx}\, dx = \frac{e^{it} - 1}{it}\,.$$

The characteristic function of a random variable η having Poisson distribution with parameter λ is given by the formula

$$\varphi_\eta(t) = \sum_{k=0}^{\infty} e^{itk}\, \frac{\lambda^k}{k!}\, e^{-\lambda} = \exp\{\lambda\,(e^{it} - 1)\}\,.$$

We list the basic properties of characteristic functions.

(1) A characteristic function $\varphi(t)$ is uniformly continuous on the real line and satisfies the conditions

$$\varphi(0) = 1\,, \qquad |\varphi(t)| \le 1\,, \qquad -\infty < t < \infty\,.$$

(2) If random variables ξ and η are such that $\eta = a\xi + b$, where a and b are arbitrary constants, then

$$\varphi_\eta(t) = e^{ibt}\, \varphi_\xi(at)\,.$$

(3) If the kth absolute moment of a random variable ξ exists, $\mathbf{E}|\xi|^k < \infty$, $k \ge 1$, then the kth derivative of $\varphi_\xi(t)$ exists and

$$\varphi^{(k)}(0) = i^k\, \mathbf{E}\,\xi^k\,.$$

In addition, in a neighborhood of the point $t = 0$, $\varphi_\xi(t)$ admits the expansion

$$\varphi_\xi(t) = 1 + \sum_{j=1}^{k} \frac{t^j i^j}{j!}\, \mathbf{E}\,\xi^j + o(t^k)\,.$$

(4) A distribution function $F(x)$ is uniquely determined by its characteristic function $\varphi(t)$, namely: for each continuity point of $F(x)$ the inversion formula

$$F(x) = \frac{1}{2\pi} \lim_{y \to -\infty} \lim_{A \to \infty} \int\limits_{-A}^{A} \frac{e^{-itx} - e^{-ity}}{it}\, \varphi(t)\, dt$$

is valid. If $p(x)$ is the density of $F(x)$ and $\varphi(t)$ is integrable then

$$p(x) = \frac{1}{2\pi} \int_{-\infty}^{\infty} e^{-itx} \varphi(t)\, dt.$$

(5) A random variable ξ is said to be *lattice* with step h if there exists a number a such that

$$\sum_{k=-\infty}^{\infty} \mathbf{P}\{\xi = a + kh\} = 1.$$

If ξ is a lattice random variable then

$$\varphi_\xi(t) = \exp\{ita\} \sum_{k=-\infty}^{\infty} \mathbf{P}\{\xi = a + kh\} \exp\{itkh\}.$$

One can show that ξ is a lattice random variable with step h if and only if

$$|\varphi_\xi(2\pi/h)| = 1.$$

If $g_\xi(x)$ is the generating function of an integer-valued random variable ξ then

$$\varphi_\xi(t) = g_\xi\!\left(e^{it}\right).$$

Putting $z = e^{it}$ and assuming that $g_\xi(z)$ has no singular points on the circle $|z| = 1$, one can write in this case

$$\mathbf{P}\{\xi = k\} = \frac{1}{2\pi i} \int_{|z|=1} g_\xi(z)\, \frac{dz}{z^{k+1}}.$$

(6) If $\xi_1, \xi_2, \ldots, \xi_n$ are mutually independent random variables and $\zeta = \xi_1 + \xi_2 + \cdots + \xi_n$ then

$$\varphi_\zeta(t) = \varphi_{\xi_1}(t)\, \varphi_{\xi_2}(t) \cdots \varphi_{\xi_n}(t).$$

(7) Let ξ be a normally distributed random variable with parameters (m, σ), that is, a random variable having density

$$(2\pi\sigma^2)^{-1/2} \exp\{-(x-m)^2/(2\sigma^2)\}, \qquad x \in (-\infty, \infty).$$

The characteristic function of ξ is

$$\varphi_\xi(t) = \exp\{imt - \tfrac{1}{2}\sigma^2 t^2\}.$$

The kth central moment of ξ is calculated by the formula

$$\mu_k = \begin{cases} \dfrac{\sigma^k\, k!}{2^{k/2}\, (k/2)!} & \text{if } k \text{ is even}, \\[2ex] 0 & \text{if } k \text{ is odd}. \end{cases}$$

1.2.2   Continuity theorem and limit theorems

Let $F(x) = \mathbf{P}\{\xi < x\}$ and $F_n(x) = \mathbf{P}\{\xi_n < x\}$, $n = 1, 2, \ldots$. The sequence of distributions $\{F_n(x)\}$ is said to converge weakly to $F(x)$ if $F_n(x) \to F(x)$ as $n \to \infty$ at each point of continuity of $F(x)$. We denote the *weak convergence* of $F_n(x)$ to $F(x)$ by

$$F_n(x) \Longrightarrow F(x).$$

The following *Lévy–Cramér theorem*, termed usually the *continuity theorem* [15, 87], plays an important role in the theory of limiting distributions.

Theorem 2.1 *Let $\{\xi_n\}$ be a sequence of random variables and let $\{\varphi_n(x)\}$ and $\{F_n(x)\}$ be the corresponding sequences of characteristic and distribution functions.*

(a) *If there exists a random variable ξ with distribution function $F(x)$ such that $F_n(x) \Rightarrow F(x)$ then*

$$\lim_{n \to \infty} \varphi_n(t) = \varphi(t)$$

uniformly over each finite interval; here $\varphi(t)$ is the characteristic function of ξ.

(b) *If there exists a function $\varphi(t)$, continuous at $t = 0$, such that*

$$\lim_{n \to \infty} \varphi_n(t) = \varphi(t)$$

for any $t \in (-\infty, \infty)$ then there exists a random variable ξ with distribution function $F(x)$ such that

$$F_n(x) \Longrightarrow F(x)$$

uniformly over any finite or infinite interval of continuity of $F(x)$. In addition, the characteristic function of ξ coincides with $\varphi(t)$ and

$$\lim_{n \to \infty} \varphi_n(t) = \varphi(t)$$

uniformly over each finite interval.

We note that the condition of continuity of $\varphi(t)$ at zero in (b) may be substituted by the condition of the uniform convergence of $\varphi_n(t)$ to $\varphi(t)$ over an arbitrary finite interval enclosing the origin.

As a corollary to the theorem we obtain the *central limit theorem* for sums of independent identically distributed random variables [15, 87]:

Theorem 2.2 *Let $\{\xi_n\}$ be a sequence of independent identically distributed random variables such that*

$$\mathbf{E}\,\xi_n = a\,, \qquad \mathrm{Var}\,\xi_n = \sigma^2\,, \qquad \zeta_n = \frac{1}{\sigma\sqrt{n}}\,(\xi_1 + \xi_2 + \cdots + \xi_n - a\,n)\,.$$

If $0 < \sigma^2 < \infty$ then

$$\lim_{n\to\infty}\mathbf{P}\{\zeta_n < x\} = \Phi(x)\,, \qquad \Phi(x) = \frac{1}{\sqrt{2\pi}}\int_{-\infty}^{x}\exp\{-\tfrac{1}{2}u^2\}\,du$$

uniformly for $x \in (-\infty,\infty)$.

In this case the sequence $\{\zeta_n\}$ is said to be *asymptotically normal* with parameters $(0,1)$.

Proof of Theorem 2.2 We may assume, with no loss of generality, that $a = 0$. In this case, due to the Lévy–Cramér theorem, it is sufficient to show that

$$\lim_{n\to\infty}\varphi_{\zeta_n}(t) = \exp\{-\tfrac{1}{2}t^2\}\,.$$

Putting $\varphi(t) = \varphi_{\xi_k}(t)$ we obtain

$$\varphi_{\zeta_n}(t) = \varphi^n\big(t/(\sigma\sqrt{n})\big)\,.$$

Property (3) of characteristic functions allows us to write the following expansion in a neighborhood of $t = 0$:

$$\varphi(t) = 1 - \tfrac{1}{2}t^2\sigma^2 + o(t^2)\,.$$

Therefore, for any fixed t, we have

$$\varphi_{\zeta_n}(t) = \left[1 - \frac{\sigma^2}{2}\left(\frac{t}{\sigma\sqrt{n}}\right)^2 + o\left(\frac{t^2}{n}\right)\right]^n\,, \qquad n \to \infty\,.$$

Hence the assertion of Theorem 2.2 follows.   □

One of the most important corollaries of the Lévy–Cramér theorem is the central limit theorem (CLT) for sums of independent random variables. We formulate the result for triangular arrays.

Let ξ_{kn}, $k = 1,2,\dots,n$; $n = 1,2,\dots$ be a triangular array of rowwise independent random variables with means $\{a_{kn}\}$ and variances $\{b_{kn}^2\}$, and let

$$B_n^2 := \sum_{k=1}^{n}b_{kn}^2\,.$$

The triangular array $\{\xi_{kn}\}$ is said to satisfy the *Lindeberg condition* if, for any $\varepsilon > 0$,

$$(2.1) \qquad \lim_{n \to \infty} \frac{1}{B_n^2} \sum_{k=1}^{n} \int_{|x-a_{kn}|>\varepsilon B_n} \left(x - a_{kn} \right)^2 dF_{kn}(x) = 0 \,,$$

where $F_{kn}(x)$ is the distribution function of the random variable ξ_{kn}.

Put

$$\zeta_n := B_n^{-1} \sum_{k=1}^{n} \left(\xi_{kn} - a_{kn} \right).$$

Theorem 2.3 *If ξ_{kn}, $k = 1, 2, \dots, n$; $n = 1, 2, \dots$, is a triangular array of rowwise independent random variables satisfying Lindeberg's condition then, as $n \to \infty$, the sequence ζ_n, $n = 1, 2, \dots$, is asymptotically normal with parameters $(0, 1)$.*

One of the versions of CLT is the *Lyapunov theorem*, which is weaker than Theorem 2.3 but whose conditions may be easier verified.

Theorem 2.4 *Let ξ_{kn}, $k = 1, 2, \dots, n$; $n = 1, 2, \dots$, be a triangular array of rowwise independent random variables and let the absolute central moments*

$$C_{kn} := \mathbf{E} \, | \, \xi_{kn} - a_{kn} \, |^{2+\delta}$$

be finite for some $\delta \in (0, 1]$. Put

$$C_n := \sum_{k=1}^{n} C_{kn}.$$

If $C_n B_n^{-1} \to 0$ as $n \to \infty$ then the corresponding sequence $\{\zeta_n\}$ is asymptotically normal with parameters $(0, 1)$.

We now derive a useful corollary to the theorem. Let us consider a triangular array of independent random variables ξ_{kn}, $k = 1, 2, \dots, n$; $n = 1, 2, \dots$, such that

$$\mathbf{P}\{\, \xi_{kn} = 1 \} = p_k \,, \qquad\qquad \mathbf{P}\{\, \xi_{kn} = 0 \} = q_k \,,$$

where $p_k = p_k(n)$, $q_k = q_k(n)$ and $p_k + q_k = 1$. We call this sequence a *Poisson sequence*. Let

$$B_n^2 = \sum_{k=1}^{n} p_k q_k \,, \qquad\qquad \eta_n = B_n^{-1} \sum_{k=1}^{n} \left(\xi_{kn} - p_k \right).$$

Corollary *If $B_n \to \infty$ as $n \to \infty$ then the sequence $\{\eta_n\}$ is asymptotically normal with parameters $(0, 1)$.*

Indeed, using the inequality $\mathbf{E}|\xi_{kn} - p_k|^3 \le p_k q_k$ and taking $\delta = 1$ in Lyapunov's theorem, we obtain $C_n \le B_n^{3/2}$. Therefore, $C_n B_n^{-1} \to 0$ as $n \to \infty$.

The limiting laws in Theorems 2.2–2.4 are described by distribution functions. Statements of this kind are called *integral limit theorems*. One more class of theorems, called *local limit theorems*, is also of great interest.

Let $\{\xi_n\}$ be a sequence of mutually independent identically distributed lattice random variables of the form

$$\xi_i = a + k\,h, \qquad\qquad k = 0,\ \pm 1,\ \pm 2,\ \dots,$$

with finite means and variances. Set

$$\zeta_n = \xi_1 + \xi_2 + \cdots + \xi_n, \qquad A_n = \mathbf{E}\,\zeta_n, \qquad B_n^2 = \operatorname{Var}\zeta_n$$

and

$$P_n(k) = \mathbf{P}\{\zeta_n = n\,a + k\,h\}, \qquad z_{nk} = B_n^{-1}(a\,n + k\,h - A_n).$$

The following local limit theorem, which is due to Gnedenko [47], is valid.

Theorem 2.5 *In order that the relation*

$$P_n(k)\,\frac{B_n}{h} - \frac{1}{\sqrt{2\pi}}\,\exp\left\{-\tfrac{1}{2}\,z_{nk}^2\right\} \longrightarrow 0, \qquad\qquad n \to \infty,$$

holds uniformly for k, $-\infty < k < \infty$, it is necessary and sufficient that the step h be taken to be maximal.

Note that a step h is maximal if there exist no $b \in (-\infty, \infty)$ and $h_1 > h$ such that all possible values of ξ_i have the form $b + k h_1$, $k = 0, \pm 1, \pm 2, \dots$.

Consider a sequence of independent identically distributed random variables ξ_n with zero mean and variance $\sigma^2 > 0$. In addition, assume the existence of a density $p_n(x)$ of the random variable $\zeta_n = \sigma^{-1} n^{-1/2} \sum_{i=1}^{n} \xi_i$.

Theorem 2.6 *In order that the relation*

$$p_n(x) - \frac{1}{\sqrt{2\pi}}\,\exp\left\{-\tfrac{1}{2}\,x^2\right\} \longrightarrow 0, \qquad\qquad n \to \infty,$$

holds uniformly for $x \in (-\infty, \infty)$ it is necessary and sufficient that the density $p_N(x)$ be bounded for some N.

Various generalizations of Theorems 2.5–2.6 are possible but we do not mention them here. We note only that if the conditions of some local limit theorem hold then the corresponding version of the integral CLT is also valid. The converse statement is, generally speaking, not true.

1.2.3 Curtiss' theorem

Let ξ be a random variable with distribution function $F(x)$. The *moment generating function* of ξ is the function of real t

$$(2.2) \qquad g(t) = \int\limits_{-\infty}^{\infty} e^{xt}\,dF(x)$$

if the integral is convergent for all t from an interval $-\delta \leq t \leq \delta$, $\delta > 0$. In the case of integer-valued random variables, this definition coincides with the definition of the moment generating function given above. If the moment generating function of a random variable exists in a neighborhood of zero then it determines the corresponding distribution uniquely. The characteristic function $\varphi(t)$ and the moment generating function $g(t)$ of one and the same distribution are related by the equality

$$\varphi(t) = g(it).$$

The moment generating function of a normal distribution with parameters (m, σ) has the form

$$g_{m,\sigma}(t) = \exp\left\{ mt + \tfrac{1}{2}\sigma^2 t^2 \right\}.$$

In the following we will often use a limit theorem proved by Curtiss [21].

Theorem 2.7 *Let $F_n(x)$ and $g_n(t)$ be the distribution function and the moment generating functions, respectively, of a random variable ξ_n. If $g_n(t)$ exists for $|t| < t_1$ and all $n > n_0$ and there exists a function $g(t)$, determined and bounded for $|t| \leq t_2 < t_1$, $t_2 > 0$, such that*

$$\lim_{n\to\infty} g_n(t) = g(t), \qquad |t| < t_2,$$

then there exists a random variable ξ with distribution function $F(x)$ such that

$$\lim_{n\to\infty} F_n(x) = F(x)$$

at all continuity points of $F(x)$, and the convergence is uniform on any finite or infinite continuity interval of $F(x)$. The moment generating function of ξ exists for $|t| \leq t_2$ and is equal to $g(t)$ in the interval.

Proof We consider the Laplace transform

$$\varphi_n(z) = \int\limits_{-\infty}^{\infty} e^{zx}\,dF_n(x).$$

Observe that if $z = t + i\alpha$ then for $n \geq n_0$ and any z from the strip $-t_1 < \operatorname{Re} z < t_1$ one obtains $|\varphi_n(z)| \leq \varphi_n(t) = g_n(t)$. Differentiating (2.2) with respect to t, where $F_n(x)$ and $g_n(t)$ are substituted for $F(x)$ and $g(t)$, respectively, we find

$$g_n''(t) = \int_{-\infty}^{\infty} x^2 e^{xt} \, dF_n(x), \qquad |t| < t_1,$$

whence it follows that $g_n''(t) > 0$ as $|t| < t_1$. This means that $g_n(t)$ attains its maximal value on the interval $|t| \leq t_2$ either at one or at both end points of the interval under consideration. Since $g_n(t_2)$ and $g_n(-t_2)$ have finite limits as $n \to \infty$, the sequence $\{g_n(t)\}$ is uniformly bounded on the interval $|t| \leq t_2$ for $n \geq n_0$. Thus, the sequence $\{|\varphi_n(z)|\}$ is uniformly bounded in the strip $-t_2 \leq \operatorname{Re} z \leq t_2$ for $n \geq n_0$ and, moreover, it has a limit at each point of any infinite set having a limiting point in the strip, that is, at each point of the interval $-t_2 \leq z \leq t_2$. By Vitali's theorem, there exists an analytic function $\varphi^*(z)$ such that

$$\lim_{n \to \infty} \varphi_n(z) = \varphi(z)$$

uniformly over any closed subdomain of the strip $-t_2 < \operatorname{Re} z < t_2$. As $\varphi_n(i\alpha)$ is the characteristic function of ξ_n, the continuity theorem provides the existence of a distribution function $F(x)$ such that $F_n(x) \Rightarrow F(x)$. In addition, $\varphi^*(t) = g(t)$, $-t_2 < t < t_2$.

It remains to prove that $\varphi^*(t)$ is the moment generating function of ξ. By the continuity theorem $\varphi^*(i\alpha)$ is the characteristic function of ξ. If we succeed in showing that $\varphi(z)$ exists at least in some strip $-t_2 < \operatorname{Re} z < t_2$ then the equality $\varphi(z) \equiv \varphi^*(z)$, which will be valid for each z belonging to the imaginary axis, will imply the validity of the same equality for each z from the strip and, in particular, for each z from the interval of the real axis lying within the strip. In order to obtain the desired statement it suffices to show that $\varphi(t)$ exists for all t, $-t_2 \leq t \leq t_2$. Let us assume that $\varphi(t)$ does not exist at an inner interval point $t = t_3$. If, under this assumption, M is the least upper bound of $g_n(t_3)$, $n \geq n_0$, there exists a real number A such that

$$\int_{-A}^{A} e^{x t_3} \, dF(x) > M.$$

On the other hand,

$$\int_{-A}^{A} e^{x t_3} \, dF(x) = \int_{-A}^{A} e^{x t_3} \, dF_n(x) + \left[\int_{-A}^{A} e^{x t_3} \, dF(x) - \int_{-A}^{A} e^{x t_3} \, dF_n(x) \right].$$

Since

$$\lim_{n \to \infty} F_n(x) = F(x)$$

at all continuity points of $F(x)$, the quantity in square brackets tends to zero, as $n \to \infty$, for each point of an everywhere dense set of the interval $[-t_2, t_2]$. Also,

$$\int_{-A}^{A} e^{x t_3} \, dF_n(x) \le \int_{-\infty}^{\infty} e^{x t_3} \, dF_n(x) \le M, \qquad n \ge n_0.$$

This contradiction implies the existence of $\varphi(t)$ for $-t_2 < t < t_2$ and, therefore, $\varphi(z)$ exists for $-t_2 < \operatorname{Re} z < t_2$. Consequently, $g(t) = \varphi^*(t) = \varphi(t)$ for $-t_2 < t < t_2$. Because the integrals determining $g_n(t)$ and $\varphi(t)$ are absolutely and uniformly convergent, these functions are continuous on the interval $-t_2 < t < t_2$. Since $\lim_{n \to \infty} g_n(t) = g(t)$ uniformly for $-t_2 < t < t_2$, the function $g(t)$ is also continuous on the interval. This means that $\varphi(t) = g(t)$ for $-t_2 < t < t_2$, which completes the proof. $\square$

If, for example, a random variable η_n has binomial distribution with parameters (n, p) then the moment generating function of the random variable $\xi_n = (\eta_n - np)(npq)^{-1/2}$, $p + q = 1$, is

$$f_n(t) = \exp\left\{ -\frac{\sqrt{np}}{q} t \right\} \left(p \exp\left\{ \sqrt{\frac{np}{q}}\, t \right\} + q \right)^n.$$

In view of the relation $\lim_{n \to \infty} f_n(t) = e^{t^2/2}$, the random variable ξ_n has asymptotically normal distribution with parameters $(0,1)$. If $p = p(n)$ and $np \to \lambda \in (0, \infty)$ as $n \to \infty$ then the moment generating function $\varphi_n(t) = \left(p e^t + q \right)^n$ of the random variable η_n has the limit $\exp\left\{ \lambda \left(e^t - 1 \right) \right\}$, that is, the limiting distribution of η_n is Poisson with parameter λ.

1.2.4 Convergence of distributions determined by convergence of moments

In subsequent chapters we will frequently use a theorem concerning the convergence of sequences of distributions given that the corresponding sequences of moments converge.

Theorem 2.8 *Let $\{F_n(x)\}$ be a sequence of distribution functions whose moments*

$$M_{kn} = \int_{-\infty}^{\infty} x^k \, dF_n(x)$$

are finite for $k = 1, 2, \ldots$, and, for each $k \ge 1$, let

$$\lim_{n \to \infty} M_{kn} = M_k \ne \pm\infty.$$

Then there exists a subsequence $\{F_{nj}(x)\}$ converging weakly to a distribution function $F(x)$ whose sequence of moments coincides with $\{M_k\}$. If the moments determine $F(x)$ uniquely then the sequence $\{F_n(x)\}$ converges weakly to $F(x)$.

It is known that a distribution can be uniquely reconstructed by the sequence $\{\mu_j\}$ of its central moments if

$$\sum_{j=0}^{\infty} \mu_{2j}^{-1/(2j)} = \infty.$$

Using this criterion we conclude that, for example, a normal distribution is uniquely determined by its moments. The same is valid for a Poisson distribution. Let $E_{n1}, \ldots, E_{nn}$ be a sequence of series of random events. Denote by $\mathbf{P}\big(E_{n,j_1} E_{n,j_2} \cdots E_{n,j_k}\big)$ the probability of the simultaneous occurrence of the events $E_{n,j_1}, E_{n,j_2}, \ldots, E_{n,j_k}$.

Corollary 1 *Let a constant $\lambda \in (0, \infty)$ exist such that, for any $k = 1, 2, \ldots,$*

$$\lim_{n \to \infty} \sum_{1 \le j_1 < \cdots < j_k \le n} \mathbf{P}\big(E_{n,j_1} E_{n,j_2} \cdots E_{n,j_k}\big) = \frac{\lambda^k}{k!},$$

where the sum is taken over all combinations of size k from the numbers $1, 2, \ldots, n$. Then the probability $P_n(r)$ of the occurrence of exactly r events in the nth series has the limit

$$\lim_{n \to \infty} P_n(r) = \frac{\lambda^r}{r!}\, e^{-\lambda}, \qquad r = 0, 1, \ldots.$$

Let $\varepsilon_{n1}, \varepsilon_{n2}, \ldots, \varepsilon_{nn}$ be random variables taking the values 0 and 1 and $\varepsilon_n := \sum_{j=1}^{n} \varepsilon_{nj}$. Denote by B_{kn} the kth binomial moment of ε_n.

Corollary 2 *If*

$$\lim_{n \to \infty} B_{kn} = \frac{\lambda^k}{k!}$$

for all $k = 1, 2, \ldots,$ then

$$\lim_{n \to \infty} \mathbf{P}\{\varepsilon_n = r\} = \frac{\lambda^r}{r!}\, e^{-\lambda}, \qquad r = 0, 1, \ldots.$$

1.3 Probability distributions in combinatorial analysis

We consider a number of combinatorial problems in which the asymptotic distributions of some or other characteristics can be found by the limit theorems formulated in the preceding section.

1.3.1 Inversions in a random permutation

Let $(a_1, a_2, \ldots, a_n)$ be a permutation of numbers $1, 2, \ldots, n$. A pair of elements (a_i, a_j) of the permutation is said to be an *inversion* if $i < j$ and $a_i > a_j$. If $B(n, r)$ is the number of permutations of n elements with r inversions then

$$B(n, r) = \sum_{s=0}^{r \wedge (n-1)} B(n - 1, r - s),$$

where we make use of the notation $a \wedge b := \min(a, b)$. Using the recurrence relation and putting $N = \binom{n}{2}$ we find the generating function of the sequence $B(n, r)$, $r = 0, 1, \ldots, N$:

$$b_n(x) = \sum_{r=0}^{N} B(n, r)\, x^r = \prod_{k=1}^{n} \frac{1 - x^k}{1 - x}.$$

If ξ_n is now the number of inversions in a permutation of n elements chosen at random and equiprobably from the set of all such permutations then the probability generating function of the random variable ξ_n has the form

$$P_n(x) = \prod_{k=1}^{n} \frac{1 - x^k}{k\,(1 - x)}.$$

Let us consider the moment generating function of ξ_n:

$$M(x; n) = P_n(e^x) = \exp\{\tfrac{1}{4} n\,(n - 1)\,x\} \prod_{j=1}^{n} \frac{\sinh(xj/2)}{j\,\sinh(x/2)}.$$

The explicit formula for the generating function of the *Bernoulli numbers* B_k, $k = 0, 1, \ldots$,

$$\frac{x}{e^x - 1} = \sum_{k=0}^{\infty} B_k \frac{x^k}{k!}, \qquad\qquad |x| < 2\pi,$$

justifies the expansion

$$\log \frac{\sinh(x/2)}{x/2} = \sum_{k=1}^{\infty} B_{2k} \frac{x^{2k}}{2k\,(2k)!}, \qquad\qquad |x| < 2\pi.$$

Using this expansion we find

$$\log M(x; n) = \frac{n\,(n - 1)}{4}\, x + \sum_{k=1}^{\infty} B_{2k} \frac{x^{2k}}{2k\,(2k)!} \sum_{j=1}^{n} (j^{2k} - 1).$$

Hence, in particular, the formulae for the mean and variance of ξ_n follow:

$$\mathbf{E}\,\xi_n = \frac{n\,(n - 1)}{4}, \qquad \mathrm{Var}\,\xi_n = \sigma^2 = \frac{2n^3 + 3n^2 - 5n}{72}.$$

Also, we have

$$(3.1) \qquad \log M\left(\frac{t}{\sigma}; n\right) = \frac{n(n-1)}{4}\frac{t}{\sigma} + \frac{t^2}{2} + R_n(t),$$

where

$$R_n(t) = \sum_{k=2}^{\infty} B_{2k} \frac{t^{2k}}{2k\,\sigma^{2k}\,(2k)!} \sum_{j=1}^{n}(j^{2k}-1).$$

Since for all $k \geq 2$

$$\sigma^{-2k} \sum_{j=1}^{n} j^{2k} = O\left(n^{1-k}\right),$$

the remainder term $R_n(t) \to 0$ as $n \to \infty$ uniformly for t from any bounded set. Observe that

$$\bar{M}(t; n) = \exp\left\{-\frac{n(n-1)}{4}\frac{t}{\sigma}\right\} M\left(\frac{t}{\sigma}; n\right)$$

is the moment generating function of the random variable

$$\eta_n = \sigma^{-1}\left(\xi_n - \tfrac{1}{4}n(n-1)\right).$$

Representation (3.1) implies

$$\lim_{n\to\infty} \bar{M}(t; n) = \exp\left\{\tfrac{1}{2}t^2\right\}$$

for any fixed t. The right-hand side of the equality is the moment generating function of the standard normal distribution. Now applying Theorem 2.7 we see that the following statement is valid.

Theorem 3.1 *If ξ_n is the number of inversions in a random equiprobable permutation of n elements then the random variable*

$$\eta_n = (\xi_n - \mathbf{E}\,\xi_n)\left(\operatorname{Var}\xi_n\right)^{-1/2}$$

has, as $n \to \infty$, asymptotically normal distribution with parameters $(0,1)$.

1.3.2 Rises in a random permutation

A pair of elements a_j and a_{j+1} in a permutation $(a_1, a_2, \ldots, a_n)$ of numbers $1, 2, \ldots, n$ is said to constitute a *rise* if $a_j < a_{j+1}$. By definition, we consider that the element a_1 is the end point of some fictitious rise (and thus the number of rises in a permutation is always positive). We denote by A_{nk} the number of permutations of n elements having r rises. These numbers, called the *Euler numbers*, satisfy the recurrence relation

$$A_{nk} = (n-k+1)\,A_{n-1,\,k-1} + k\,A_{n-1,\,k}.$$

Using this formula and the initial conditions $A_{n1} = A_{nn} = 1$, one can obtain an explicit expression for Euler's numbers:

$$(3.2) \qquad A_{nk} = \sum_{j=0}^{k} (-1)^j \binom{n+1}{j} (k-j)^n, \qquad k = 1, 2, \ldots, n.$$

Let η_n be the number of rises in a permutation chosen at random and equiprobably from the set of all permutations of n elements. The distribution of the random variable η_n can easily be recovered from (3.2):

$$(3.3) \quad \mathbf{P}\{\eta_n = k\} = \frac{1}{n!} \sum_{j=0}^{k} (-1)^j \binom{n+1}{j}(k-j)^n, \quad k = 1, 2, \ldots, n.$$

It follows from (1.29) that if random variables $\xi_1, \xi_2, \ldots, \xi_n$ are independent and each of them is uniformly distributed on the interval $[0, 1]$, then the distribution of their sum has the form

$$(3.4) \quad F_n(x) = \mathbf{P}\{\xi_1 + \xi_2 + \cdots + \xi_n < x\} = \frac{1}{n!} \sum_{j=0}^{[x]} (-1)^j \binom{n}{j}(x-j)^n,$$

where $[x]$ is the integral part of x. Relations (3.3) and (3.4) imply

$$\mathbf{P}\{\eta_n = k\} = F_n(k) - F_n(k-1), \qquad k = 1, 2, \ldots, n,$$

that is, the values of the distribution function of η_n at the points $x = 1, 2, \ldots, n$ are equal to $F_n(x)$.

To find the limiting distribution of the sum $\sum_{i=1}^{n} \xi_i$ we apply Theorem 2.2. Observe that

$$\mathbf{E}\,\xi_i = \tfrac{1}{2}, \qquad \mathrm{Var}\,\xi_i = \tfrac{1}{12}, \qquad i = 1, 2, \ldots, n,$$

and, therefore, $B_n = \sqrt{n/12}$. Thus, the random variable

$$\Big(\sum_{i=1}^{n} \xi_i - n/2 \Big) / \sqrt{n/12}$$

has, as $n \to \infty$, asymptotically normal distribution with parameters $(0, 1)$.

As the distributions of η_n and $\sum_{i=1}^{n} \xi_i$ coincide at the points $1, 2, \ldots, n$, one obtains the following well-known theorem.

Theorem 3.2 *The distribution of the random variable*

$$(\eta_n - n/2) / \sqrt{n/12}$$

converges to the standard normal distribution as $n \to \infty$.

The method described for proving the theorem was suggested by Tanny [117], who gave a probabilistic interpretation of Euler's numbers. He also proved the following version of the local limit theorem: if $x_n = x\sqrt{n/12} + n/2$ then

$$(3.5) \qquad \lim_{n \to \infty} \sqrt{\frac{n}{12}}\, \frac{A_{n,[x_n]}}{n!} = \frac{1}{\sqrt{2\pi}} \exp\{-\tfrac{1}{2}x^2\}.$$

1.3.3 A limit theorem

Let $\{a_n(k)\}$ be a sequence of nonnegative numbers and let

$$(3.6) \qquad f(z,w) := \sum_{n,\,k=0}^{\infty} a_n(k)\, z^n\, w^k$$

be the corresponding double generating function. We consider a random variable ξ_n having the probability distribution

$$(3.7) \qquad P_n(k) = a_n(k)\, A_n^{-1}, \qquad\qquad k = 0, 1, \dots,$$

where the denominator $A_n := \sum_{j=0}^{\infty} a_n(j)$ is assumed to be positive and finite. We prove the following theorem which is due to Bender [5].

Theorem 3.3 *Suppose there exist a function $A(s)$, continuous and non-zero in a neighborhood of the point 0, a function $r(s)$ with a bounded third derivative near zero, a nonnegative integer m and positive numbers ε and δ such that the function*

$$(3.8) \qquad \left(1 - \frac{z}{r(s)}\right)^m f(z, e^s) - A(s)\left(1 - \frac{z}{r(s)}\right)^{-1}$$

is analytic and bounded for $|s| < \varepsilon$ and $|z| < |r(0)| + \varepsilon$. Further, let

$$(3.9) \qquad \mu = -\frac{r'(0)}{r(0)}, \qquad \sigma^2 = \mu^2 - \frac{r''(0)}{r(0)} \neq 0.$$

As $n \to \infty$, the random variable

$$\eta_n = \left(\xi_n - \mu n\right)\left(\sigma^2 n\right)^{-1/2}$$

is asymptotically normal with parameters $(0,1)$, that is,

$$(3.10) \qquad \lim_{n\to\infty} \mathbf{P}\{\eta_n < x\} = \frac{1}{\sqrt{2\pi}} \int_{-\infty}^{x} \exp\left\{-\tfrac{1}{2} t^2\right\} dt.$$

Proof Let

$$f(z, e^s) := \sum_{n=0}^{\infty} \varphi_n(s)\, z^n.$$

Denoting by $\mathrm{coef}_{z^n} \psi(z)$ the coefficient of z^n in the expansion of $\psi(z)$ in powers of z in a neighborhood of the point 0 and using Cauchy's integral formula with the contour $|z| = |r(0)| + \tfrac{2}{3}\delta$, we obtain

$$\left| \mathrm{coef}_{z^n}\left[\left(1 - \frac{z}{r(s)}\right)^m f(z, e^s) - \frac{A(s)}{1 - z\, r^{-1}(s)} \right] \right| \leq C\left(|r(0)| + \tfrac{2}{3}\delta\right)^{-n},$$

where C is an absolute constant. Using the estimate we find

$$\left| \operatorname{coef}_{z^n} \left[f(z, e^s) - \frac{A(s)}{\left(1 - z\, r^{-1}(s)\right)^{m+1}} \right] \right|$$

$$\le C \sum_{k=0}^{n} \binom{m+k-1}{k} \left(|r(0)| + \tfrac{2}{3}\delta \right)^{-n+k} |r(s)|^{-k}.$$

Choose an s such that

$$|r(s) - r(0)| \le \tfrac{1}{3}\delta.$$

Then

$$\left| \varphi_n(t) - A(s) \binom{n+m}{m} \frac{1}{r^n(s)} \right|$$

$$\le \frac{C}{|r(s)|^n} \binom{n+m-1}{n} \sum_{k=0}^{n} \left(1 + \frac{\delta}{3\,|r(s)|} \right)^{-n+k}.$$

Since the sum is absolutely bounded for small s, it follows that

$$(3.11) \qquad \varphi_n(t) = \binom{n+m}{m} \frac{A(s)}{r^n(s)} \left(1 + O\left(n^{-1} \right) \right), \qquad n \to \infty,$$

uniformly for s from a neighborhood of the origin.

Let ξ_n be a random variable having distribution (3.7) and

$$\eta_n = (\xi_n - \mu_n)\,\sigma_n^{-1},$$

where $\mu_n = n\,\mu$, and $\sigma_n^2 = n\sigma^2$. We note that the characteristic function of η_n is of the form

$$f_n(t) = \frac{1}{\varphi_n(0)}\, \varphi_n\!\left(\frac{it}{\sigma_n} \right)\, \exp\left\{ -\frac{i\mu_n t}{\sigma_n} \right\}.$$

Because of continuity of $A(s)$ in a neighborhood of $s = 0$, equality (3.11) implies

$$(3.12) \qquad f_n(t) = \exp\left\{ -\frac{i\mu_n t}{\sigma_n} \right\} \left[\frac{r(0)}{r(it/\sigma_n)} \right]^n (1 + o(1)),$$

where $o(1) \to 0$ uniformly for t from any bounded interval. By expanding the function $\rho(s) = \log\left[r(s)/r(0) \right]$ in a Taylor series near the point $s = 0$, we obtain

$$(3.13) \qquad \rho\!\left(\frac{it}{\sigma_n} \right) = \frac{r'(0)}{r(0)}\frac{it}{\sigma_n} - \left[\frac{r''(0)}{r(0)} - \left(\frac{r'(0)}{r(0)} \right)^2 \right] \frac{t^2}{2\sigma_n^2} + o(1).$$

From (3.12) and (3.13) we know that

$$\lim_{n \to \infty} f_n(t) = \exp\left\{-\tfrac{1}{2}t^2\right\}, \qquad t \in (-\infty, \infty).$$

Now the Lévy–Cramér theorem yields Theorem 3.3.   □

One of the main advantages of the theorem is the generality of its conditions and, therefore, as one might expect, a wide range of applications in searching for limiting distributions in combinatorial problems. Unfortunately, in some cases the problem of verifying the fulfilment of these conditions is very difficult in itself. And so, in subsequent chapters, we will also use other methods for finding limiting distributions in combinatorial problems. Nevertheless, there are some problems in which asymptotic distributions may be obtained by direct application of Theorem 3.3. We consider one of them.

1.3.4 Morgan numbers

Defining, as usual, the finite difference of a function $f(x)$ at the point x as

$$\Delta f(x) := f(x+1) - f(x),$$

it is not difficult to check that

(3.14)
$$\Delta^k f(x) = \sum_{j=0}^{k} (-1)^j \binom{k}{j} f(x+k-j).$$

From (3.14), setting $\Delta^k 0^n = \Delta^k x^n \big|_{x=0}$, we find

$$\Delta^k 0^n = \sum_{j=0}^{k} (-1)^j \binom{k}{j} (k-j)^n.$$

The numbers $\Delta^k 0^n$, $k = 0, 1, \ldots, n$, are called the *Morgan numbers*. The double generating function of Morgan's numbers has the form

$$f(z, w) = \sum_{n=0}^{\infty} \sum_{k=0}^{n} \Delta^k 0^n \, w^k \, z^n = \frac{1}{1 - w(e^z - 1)}.$$

There is a simple pole of $f(z, e^s)$ at the point $z = r(s) = \log\left(1 + e^{-s}\right)$. Consequently,

$$f(z, e^s) = \frac{1}{1 + e^s} \frac{1}{1 - e^{z - r(s)}} = \frac{1}{1 + e^s} \left[\frac{1}{z - r(s)} + O(1) \right]$$

provided that $|s|$ and $|z - r(s)|$ are bounded. Hence it follows that the conditions of Theorem 3.3 hold with

$$\mu = \frac{1}{2\log 2}, \qquad \sigma = \frac{\sqrt{1 - \log 2}}{2\log 2}.$$

We consider a random variable ξ_n having the probability distribution

$$P_n(k) = \frac{\Delta^k 0^n}{\sum\limits_{k=0}^{n} \Delta^k 0^n}, \qquad k = 0, 1, \dots, n,$$

and put $\mu_n = n\mu$ and $\sigma_n^2 = n\sigma^2$. Then, according to Theorem 3.3, the random variable $(\xi_n - \mu_n)\sigma_n^{-1}$ is asymptotically normal with parameters $(0, 1)$ as $n \to \infty$.

1.4   Asymptotic formulae and limiting distributions

1.4.1   Asymptotic formulae

We consider a general method for finding asymptotic formulae and limiting distributions in combinatorial analysis. This method is a modification of the saddle point method that is used, in particular, to find an asymptotic formula, as $n \to \infty$, for the functions $Q_n(x)$ whose generating function is of the form

$$\exp\{g(x, t)\} = \sum_{n=0}^{\infty} Q_n(x) \frac{t^n}{n!},$$

where $g(x, t)$ is a function of real variables x and t. Here we restrict ourselves to the case $g(x, t) = x\,g(t)$, where $g(t) = g_m(t)$ is a polynomial of degree m in t. It will be seen in subsequent chapters that the main features of the method remain for other classes of functions $g(x, t)$.

Let $g(t) = g_m(t)$ be a polynomial with real nonnegative coefficients:

$$(4.1) \qquad g(t) := g_m(t) = \sum_{k=0}^{m} a_k\, t^k, \qquad a_m \neq 0,$$

and let

$$(4.2) \qquad \exp\{xg(t)\} := \sum_{n=0}^{\infty} Q_{nm}(x) \frac{t^n}{n!}.$$

In what follows we will assume that, in the representation

$$(4.3) \qquad g(t) = a_0 + a_r\, t^r + a_h\, t^h + \cdots + a_s\, t^s + a_m\, t^m,$$

the coefficients $a_r, a_h, \dots, a_s, a_m$ are positive and the numbers $r, h, \dots, m$ are coprime, that is,

$$(4.4) \qquad \gcd(r, h, \dots, s, m) = 1.$$

This should cause no essential loss of generality because the case

$$\gcd\left(r, h, \ldots, s, m\right) = q > 1$$

is reduced by the substitution $y = t^q$ to a generating function of the form
(4.2) with the desired property.

One of the main results of this section is a theorem giving, as $n \to \infty$,
an asymptotic representation for the coefficients $Q_{nm}(x)$ of a generating
function of the form (4.2).

Theorem 4.1 *For any fixed polynomial $g(t)$ satisfying (4.3) and (4.4), the
following asymptotic relation is valid as $n \to \infty$:*

$$(4.5) \qquad Q_{nm}(x) = \frac{n! \; \exp\left\{xg(R)\right\}}{R^n \; \sqrt{2 \, \pi \, x \, \chi^2 \, g(R)}} \left(1 + o(1)\right),$$

where χ is the operator

$$\chi := R \, \frac{d}{dR} \,,$$

R is a function of n, given by the relation

$$(4.6) \qquad \chi \, g(R) = \frac{n}{x}$$

*and uniquely determined for large n, and $o(1) \to 0$ uniformly for x from a
neighborhood $W = [1 - \delta, \, 1 + \delta]$, $0 < \delta < 1$, of the point $x = 1$.*

Proof Viewing $\exp\left\{xg(t)\right\}$ as a function of the complex variable z and
applying Cauchy's integral formula, we can write

$$(4.7) \qquad Q_{nm}(x) = \frac{n!}{2\pi i} \oint_C \exp\left\{xg(z)\right\} \frac{dz}{z^{n+1}} \,.$$

As $\exp\left\{xg(t)\right\}$ is an entire function, we may choose for C a circle of arbitrary
radius R with center at the origin. From (4.7), presenting the circle as

$$z = R \, e^{i\theta} \,, \qquad\qquad -\pi \le \theta \le \pi \,,$$

we find

$$(4.8) \qquad Q_{nm}(x) = A_{nm}(x) \, J_{nm}(x) \,,$$

where

$$(4.9) \qquad A_{nm}(x) := \frac{n!}{2\pi R^n} \, \exp\left\{xg(R)\right\} \,,$$

$$(4.10) \qquad J_{nm}(x) := \int_{-\pi}^{\pi} \exp\left\{F(R, \theta)\right\} d\theta$$

and

$$(4.11) \qquad F(R, \theta) := x \left[g(R e^{i\theta}) - g(R) \right] - i n \theta.$$

We assume that $R = R_{nm}(x)$, $x \in W$, tends to infinity as $n \to \infty$. The exact order of growth of R will be shown below while estimating the integral $J = J_{nm}(x)$. To evaluate the integral we choose

$$(4.12) \qquad \varepsilon = R^{(1-4m)/8}$$

and partition J into three integrals:

$$(4.13) \qquad J = J_1 + J_2 + J_3 = \int_{-\pi}^{-\varepsilon} + \int_{-\varepsilon}^{\varepsilon} + \int_{\varepsilon}^{\pi}.$$

It turns out that, as $n \to \infty$, the integral J_2 gives the main contribution to J, while the integrals J_1 and J_3 are small in comparison with J_2. We obtain the desired estimates for J_1 and J_3 by proving the following two lemmas.

Lemma 1 *If condition* (4.4) *holds and*

$$(4.14) \qquad \cos r\theta = 1, \quad \cos h\theta = 1, \quad \ldots \quad, \quad \cos m\theta = 1$$

for some $\theta \in [0, \pi]$ then $\theta = 0$.

Proof Assuming that (4.14) is valid for some $\theta \in (0, \pi]$, we see that

$$r\theta = 2\pi a, \quad h\theta = 2\pi b, \quad \ldots \quad , m\theta = 2\pi c$$

for some integral numbers $a, b, \ldots, c$; by condition (4.4), not all the numbers $2a/r$, $2b/h$, $\ldots$, $2c/m$ are integral. Without loss of generality, let $2a/r$ be nonintegral and $2a/r = p/q$ (here p, q are integral numbers, $q > 1$ and $\gcd(p, q) = 1$). Now

$$r\theta = r p\pi / q, \quad h\theta = h p\pi / q, \quad \ldots \quad , \quad m\theta = m p\pi / q.$$

Hence $q > 1$ divides $r, h, \ldots, m$, which contradicts the conditions of the lemma. $\square$

With the polynomial $g(t)$ defined by (4.3) we associate the trigonometric polynomial

$$(4.15) \qquad S(R, \theta) = \sum_{k=0}^{m} a_k R^k \cos k\theta,$$

which is called the *dominant function* of $g(t)$. It is worth noting that

$$S(r, \theta) = \tfrac{1}{2}\left[g(Re^{i\theta}) + g(Re^{-i\theta})\right].$$

Lemma 2 *Let* $\varepsilon = R^{(1-4m)/8}$ *and* $\varepsilon \le \theta \le \pi$. *Then, for sufficiently large* R, *we have*

(4.16) $S(R, \theta) \le S(R, \varepsilon).$

Proof In view of the conditions $a_k \ge 0$, $k = 1, 2, \ldots, m$, it follows from Lemma 1 that $S(R, \theta)$ attains its maximum value at the point $\theta = 0$. Using (4.15) we can write

$$S(R, 0) - S(R, \varepsilon) = \sum_{k=1}^{m} a_k R^k (1 - \cos k\varepsilon).$$

From this equality we derive the estimate

(4.17) $\quad S(R, 0) - S(R, \varepsilon) = \tfrac{1}{2} m^2 a_m R^{1/4} \left[1 + O\left(R^{1/4 - m}\right) + O\left(R^{-1}\right)\right].$

On the other hand, condition (4.4) and Lemma 1 imply, for any fixed θ, $\varepsilon \le \theta \le \pi$, the existence of an integer $\mu \le m$ such that $\cos \mu\theta \ne 1$ and $a_\mu \ne 0$. Hence, for any $\theta \in [\varepsilon, \pi]$, the following estimates are true:

(4.18) $S(R, 0) - S(R, \theta) = \sum_{k=1}^{m} a_k R^k (1 - \cos k\theta) \ge K_1 R^\mu \ge K_1,$

where K_1 is a positive constant. By comparing estimates (4.17) and (4.18) we see that Lemma 2 is valid.   □

First we estimate the integral J_2.

Lemma 3 *Let* $n \to \infty$ *and let* R *be the unique (for large n) real positive solution of equation* (4.6). *Then the following asymptotic representation is valid:*

(4.19) $J_2 = \sqrt{\dfrac{2\pi}{x\chi^2 g(R)}} \left(1 + o(1)\right),$

where $o(1) \to 0$ *as* $n \to \infty$ *uniformly for* $x \in W$.

Proof We expand $F(R, \theta)$ in a Maclaurin series. For $-\varepsilon \le \theta \le \varepsilon$ we write

(4.20) $F(R, \theta) = i\theta\left(x C_1(R) - n\right) + x \sum_{j=2}^{\infty} C_j(R) \dfrac{(i\theta)^j}{j!},$

where

(4.21) $$C_j(R) := \chi^j g(R) = \sum_{k=1}^{m} k^j a_k R^k, \qquad j = 1, 2, \dots .$$

In view of (4.6) we have $C_1(R) = n/x$ and, therefore, for each $x \in W$ and sufficiently large n, R is the unique positive solution of the equation

(4.22) $$\sum_{k=1}^{m} k a_k R^k = \frac{n}{x}$$

and it may be calculated by iteration starting with

$$R \sim \left(\frac{n}{x \, m \, a_m} \right)^{1/m} .$$

Equality (4.20) can be written in the form

(4.23) $$F(R, \theta) = -\tfrac{1}{2} x C_2(R) \theta^2 + \Phi(R, \theta) ,$$

where

(4.24) $$\Phi(R, \theta) = x \sum_{j=3}^{\infty} C_j(R) \frac{(i \theta)^j}{j!} .$$

Using (4.12), (4.21) and (4.24) it is not difficult to check that

(4.25) $$\Phi(R, \theta) = o(1) ,$$

where $o(1) \to 0$ uniformly for $x \in W$.

We make a substitution in the integral J_2 putting

$$\varphi = \theta \sqrt{\tfrac{1}{2} x C_2(R)} .$$

Taking (4.24) and (4.25) into account, we obtain

(4.26) $$J_2 = \sqrt{\frac{2}{x \, C_2(R)}} \int_{-\gamma}^{\gamma} e^{-\varphi^2} \, d\varphi \left(1 + o(1) \right) ,$$

where $H_1 R^{1/8} < \gamma < H_2 R^{1/8}$ for some positive constants H_1 and H_2, and $o(1) \to 0$ uniformly for $x \in W$.

From representation (4.26), using the equality

$$\int_{-\infty}^{\infty} e^{-\varphi^2} \, d\varphi = \sqrt{\pi} ,$$

we deduce that

(4.27) $$J_2 = \sqrt{\frac{2 \pi}{x \chi^2 g(R)}} \left(1 + o(1) \right) ,$$

where $o(1) \to 0$ uniformly for $x \in W$. $\qquad \Box$

Our next step is to estimate the integrals J_1 and J_3.

Lemma 4 *As* $n \to \infty$, *the asymptotic estimates*

$$|J_l| = o(1), \qquad l = 1, 3,$$

are valid, where $o(1)$ *decreases exponentially and* $o(1) \to 0$ *uniformly for* $x \in W$.

Proof Equalities (4.10), (4.11) and (4.13) imply that

$$|J_l| \leq \int_\varepsilon^\pi \exp\left\{ x\left[S(R, \theta) - S(R, 0) \right] \right\} d\theta, \qquad l = 1, 3.$$

Further, applying estimate (4.17) and Lemma 2, we obtain

$$|J_l| = O\left(\exp\left\{ -C_l R^{1/4} \right\} \right), \qquad l = 1, 3,$$

where the C_l are positive constants independent of x. This completes the proof of the lemma. $\qquad\square$

Now Theorem 4.1 follows from (4.8)–(4.9) and Lemmas 3–4. $\qquad\square$

Corollary ([81–82]) *Let* $g(t)$ *be a polynomial of the form* (4.2) *and let conditions* (4.3)–(4.4) *hold and*

$$(4.28) \qquad \exp\left\{ g(t) \right\} = \sum_{n=0}^{\infty} Q_{nm} \frac{t^n}{n!}.$$

Then, as $n \to \infty$, *the following asymptotic representation is valid:*

$$(4.29) \qquad Q_{mn} = \frac{n! \, \exp\left\{ g(r) \right\}}{r^n \sqrt{2\pi \chi^2 g(r)}} \left(1 + o(1) \right),$$

where, for large n, r *is the unique positive solution of the equation*

$$(4.30) \qquad \chi g(r) = C_1(r) = n.$$

The corollary follows from Theorem 4.1 by setting $x = 1$. $\qquad\square$

We now show applications of the results obtained.

Example 1 Let $m = 1$ and $g(t) = t$. In this case $Q_{n1} = 1$, $n = 0, 1, \ldots$, equation (4.22) takes the form $r = n$ and $C_j(r) = r = n$, $j > 1$. Applying (4.29) yields the well-known Stirling expansion for $n!$:

$$(4.31) \qquad n! = \sqrt{2\pi n} \, (n/e)^n \left(1 + o(1) \right).$$

Example 2 Let d be a positive integer, $\mathbb{S}_n$ be the symmetric group of order n and let $\mathbf{e}$ be the identity permutation in $\mathbb{S}_n$. A solution $\mathbf{s} \in \mathbb{S}_n$ of the equation

$$(4.32) \qquad \mathbf{s}^d = \mathbf{e}$$

is a permutation with cycles whose lengths are divisors of d. It is proved in [80] that if A_{nd} is the number of solutions of (4.32) then

$$\sum_{n=0}^{\infty} A_{nd} \frac{t^n}{n!} = \exp\left\{ \sum_{k \mid d} \frac{t^k}{k} \right\}.$$

In particular, if $d = p$, where p is a prime, then

$$(4.33) \qquad \sum_{n=0}^{\infty} A_{np} \frac{t^n}{n!} = \exp\left\{ t + p^{-1} t^p \right\}.$$

In this case $g(t) = t + p^{-1} t^p$, $\chi^2 g(r) = r + p r^p$ and equation (4.3) takes the form

$$(4.34) \qquad r^p + r = n.$$

Applying (4.29) we have

$$(4.35) \qquad A_{np} = \frac{n!\, \exp\left\{ r + p^{-1} r^p \right\}}{r^n \left[2\pi \left(r + p r^p \right) \right]^{1/2}} \left(1 + o(1) \right).$$

Using equation (4.34) we obtain

$$r^n = n^{n/p} \exp\left\{ -\frac{r}{p} - \frac{r^2}{2np} \right\} \left(1 + o(1) \right),$$

$$(r + p r^p)^{1/2} = (np)^{1/2} \left(1 + o(1) \right).$$

From (4.35), combining the preceding formulae and expanding $n!$ by Stirling's formula, we derive the representation

$$A_{np} = \frac{1}{\sqrt{p}} \left(\frac{n}{e} \right)^{n(1 - 1/p)} \exp\left\{ r + \frac{r^2}{2np} \right\} \left(1 + o(1) \right).$$

For $p = 2$ we have

$$r + \frac{r^2}{4n} = \sqrt{n} - \frac{1}{4} + o(1)$$

and, therefore,

$$(4.36) \qquad A_{n2} = \frac{1}{\sqrt{2}} \left(\frac{n}{e}\right)^{n/2} \exp\left\{\sqrt{n} - \frac{1}{4}\right\} \left(1 + o(1)\right).$$

If $p > 2$ then

$$r + \frac{r^2}{2\,n\,p} = n^{1/p} + o(1)$$

and, consequently,

$$(4.37) \qquad A_{np} = \frac{1}{\sqrt{p}} \left(\frac{n}{e}\right)^{n\,(1-1/p)} \exp\left\{n^{1/p}\right\} \left(1 + o(1)\right).$$

We note in conclusion that, generally speaking, the condition for the coefficients of $g(t)$ may not be weakened. The reason is that, under the conditions $a_k \geq 0$, $k = 1, 2, \ldots, m$, the dominant function $S(R, \theta)$ of $g(t)$ attains its maximum at the point $\theta = 0$. However, if we allow the coefficients to be negative then $S(R, \theta)$ may attain its maximum at some point $\theta \neq 0$, which depends on R, and it may arise that the number of such points is more than one.

1.4.2 Limiting distributions

We consider a sequence of integer-valued random variables $\{\xi_n\}$ and assume that, for each $n = 1, 2, \ldots$, the random variable ξ_n has the moment generating function and, as $n \to \infty$, the generating function of ξ_n has an asymptotic representation of the form

$$(4.38) \qquad f_n(t) = x^{v_n} \exp\left\{h_n(x)\right\} \left(1 + o(1)\right).$$

Here $\{v_n\}$ is a sequence of real numbers, the function $h_n(x)$ is, for each $n = 1, 2, \ldots$, three times differentiable in a neighborhood $W = [1 - \delta, 1 + \delta]$, $\delta > 0$, of the point $x = 1$ and its derivatives (up to the third order) are continuous in W, and $o(1)$ tends to 0, as $n \to \infty$, uniformly for $x \in W$. Observe that $h_n(1) \to 0$ as $n \to \infty$. Denoting differentiation with respect to x by primes, we formulate the following statement.

Theorem 4.2 *If, as $n \to \infty$, the generating function $f_n(x)$ of a random variable ξ_n has a representation of the form (4.38) and*

$$(4.39) \qquad \frac{h'''(x)}{\left(h_n'(1) + h_n''(1)\right)^{3/2}} \longrightarrow 0$$

uniformly for $x \in W$, then the random variable

$$\eta_n = \frac{\xi_n - v_n - h_n'(1)}{\left(h_n'(1) + h_n''(1)\right)^{1/2}}$$

has asymptotically normal distribution with parameters $(0, 1)$ as $n \to \infty$.

Proof We expand $h_n(x)$ by Taylor's formula in a neighborhood W of the point $x = 1$:

$$(4.40) \qquad h_n(x) = h_n(1) + h_n'(1)(x-1) + \tfrac{1}{2} h_n''(1)(x-1)^2$$

$$+ \tfrac{1}{6} h_n'''(1 + \theta(x-1))(x-1)^3.$$

Setting $\sigma^2 = h_n'(1) + h_n''(1)$, we substitute t for x by the equality $x = \exp\{t/\sigma\}$. The neighborhood W of $x = 1$ corresponds, under this substitution of variables, to a neighborhood W' of $t = 0$. In the last neighborhood the following expansion is valid:

$$h_n(e^{t/\sigma}) = h_n'(1)\,\frac{t}{\sigma} + \left[h_n'(1) + h_n''(1)\right]\frac{t^2}{2\sigma^2} + o(1).$$

For the moment generating function of the random variable η_n,

$$(4.41) \qquad \varphi_n(t) = \exp\left\{ -(v_n + h_n'(1))\,\frac{t}{\sigma} \right\} f_n(e^{t/\sigma}),$$

we know the representation

$$\varphi_n(t) = \exp\left\{ -h_n'(1)\,\frac{t}{\sigma} + h_n(e^{t/\sigma}) \right\} (1 + o(1)).$$

Hence we conclude that

$$(4.42) \qquad \lim_{n\to\infty} \varphi_n(t) = e^{t^2/2}$$

for any $t \in W'$. Now the validity of the theorem follows from Curtiss' theorem. $\qquad \square$

If, for example, ξ_n is the number of successes in a sequence of n independent trials, each of which is successful with probability p and fails with probability q, $p + q = 1$, then the generating function $f_n(x)$ of ξ_n is calculated by the formula $f_n(x) = (px + q)^n$. In this case,

$$v_n = 0, \qquad h_n(x) = n\log(px + q), \qquad h_n'(1) = np,$$

$$h_n'(1) + h_n''(1) = npq, \qquad h_n'''(x) = \frac{2np^3}{(px + q)^3}$$

and condition (4.39) holds. Hence the well-known fact follows that the random variable $(\xi_n - np)/\sqrt{npq}$ is asymptotically normal with parameters $(0, 1)$.

Consider a sequence of random variables ξ_n whose generating functions $f_{mn}(x)$ are of the form

$$(4.43) \qquad f_{nm}(x) = Q_{nm}(x)/Q_{nm}(1), \qquad n = 0, 1, \dots,$$

where $Q_{nm}(x)$ are defined by (4.2) with a polynomial $g(t)$ of degree m satisfying conditions (4.3) and (4.4). For $s = \max\{j < m : a_j \neq 0\}$ we put

(4.44) $$U(R) := \sum_{k=0}^{s} a_k R^k, \qquad\qquad s \geq 1,$$

(4.45) $$U_j(R) := \chi^j\, U(R), \qquad\qquad j = 1, 2, \dots,$$

where, as before,

$$\chi := t\,\frac{d}{dt}$$

and R is, for large n, the unique real positive solution of (4.6).

Let us introduce the notations:

(4.46) $$M_n := U(r) - \frac{C_1(r)}{C_2(r)}\, U_1(r)$$

$$+ \frac{1}{m}\left(U_1(r) - \frac{C_1(r)}{C_2(r)}\, U_2(r)\right) \frac{U_1(r)}{n - U_1(r)}\,,$$

(4.47) $$\sigma_n^2 := U(r) - \frac{2}{m}\, U_1(r) + \frac{1}{m^2}\, U_2(r)\,,$$

where the $C_j(r)$, $j = 1, 2, \dots$, are determined by (4.21) and r is, for large n, the unique positive root of equation (4.30).

We now establish a theorem by a method which will often be used below as a powerful instrument for proving limit theorems.

Theorem 4.3 *If $\{\xi_n\}$ is a sequence of random variables whose generating functions $\{f_{nm}(x)\}$ are of the form (4.43) and the corresponding polynomial $g(t)$ has $s \geq 1$ then, in the notations above, the random variable*

$$\eta_n = \left(\xi_n - n\,m^{-1} - M_n\right)\sigma_n^{-1}$$

has asymptotically normal distribution with parameters $(0,1)$ as $n \to \infty$.

Proof By Theorem 4.1 we have as $n \to \infty$

(4.48) $$f_{nm}(x) = \exp\left\{ x\,g(R) - g(r) + n\,\log\,(r/R)\right\}$$

$$\times \exp\left\{ \tfrac{1}{2}\log\left[C_2(r)\,/\,(x\,C_2(R))\right]\right\}\left(1 + o(1)\right),$$

where $o(1) \to 0$ uniformly for $x \in W$. Using equalities (4.44) and (4.45) we find that

(4.49) $$f_{nm}(x) = x^{n/m}\,\exp\left\{h_n(x)\right\}\left(1 + o(1)\right),$$

where $o(1) \to 0$ uniformly for $x \in W$, and that

$$(4.50) \qquad h_n(x) = x \left(U(R) - \frac{1}{m} U_1(R) \right)$$
$$- \left(U(r) - \frac{1}{m} U_1(r) \right)$$
$$+ \frac{n}{m} \log \frac{1 - n^{-1} U_1(r)}{1 - n^{-1} x U_1(R)}$$
$$+ \frac{1}{2} \log \frac{C_2(r)}{x C_2(R)} .$$

Taking equality (4.49) into account, we apply Theorem 4.2 with $v_n = n\,m^{-1}$ to prove Theorem 4.3. From (4.50) it follows that

$$(4.51) \qquad h'_n(1) = M_n - \frac{1}{2} \left(1 - \frac{C_1(r)\,C_3(r)}{C_2^2(r)} \right)$$

and, as $n \to \infty$,

$$(4.52)\ h'_n(1) + h''_n(1) = \left[U(r) - \frac{C_1(r)}{C_2(r)} \left(1 + \frac{C_1(r)\,C_3(r)}{C_2^2(r)} \right) \right.$$
$$\left. \times\ U_1(r) + \left(\frac{C_1(r)}{C_2(r)} \right)^2 U_2(r) \right] \left(1 + O\!\left(\frac{r^s}{n} \right) \right).$$

Hence we derive the estimates

$$(4.53) \qquad h'_n(1) = M_n + o(1) ,$$

$$(4.54) \qquad h'_n(1) + h''_n(1) = \sigma_n^2 \left(1 + o(1) \right) ,$$

where

$$(4.55) \qquad M_n = \sum_{k=1}^{s} \left(1 - \frac{k}{m} \right) a_k\, r^k \left(1 + o(1) \right) ,$$

$$(4.56) \qquad \sigma_n^2 = \sum_{k=1}^{s} \left(1 - \frac{k}{m} \right)^2 a_k\, r^k \left(1 + o(1) \right) .$$

Calculating $h'''_n(x)$ it is not difficult to check that the quantity $r^{-3}\,h'''_n(x)$ is uniformly bounded for $x \in W$. Combining this fact with formulae (4.53)–(4.56), we see that condition (4.39) of Theorem 4.2 holds. This proves Theorem 4.3. □

For small m one can find with the aid of (4.8) an asymptotic representation for $f_{nm}(x)$ with explicit dependence on x only. For example, for $m = 3$ and $g(t) = t^3 + t + 1$

$$C_1(R) = 3\,R^3 + R\,, \qquad\qquad C_2(R) = 9\,R^3 + R\,.$$

Equation (4.6) takes the form

$$3\,R^3 + R\ =\ \frac{n}{x}\ .$$

Hence, by iteration we find

$$(4.57)\qquad R = \left(\frac{n}{3\,x}\right)^{1/3}\left[1 - \frac{1}{3^{4/3}}\left(\frac{x}{n}\right)^{2/3} - \frac{1}{3^{8/3}}\left(\frac{x}{n}\right)^{4/3}\right.$$

$$\left. + \frac{1}{3^{7/3}}\left(\frac{x}{n}\right)^{5/3} + O\!\left(\frac{1}{n^2}\right)\right].$$

Using this representation it is easy to show that, as $n \to \infty$,

$$x\,g(R) - g(r)\ =\ \frac{2}{3}\left(\frac{n}{3}\right)^{1/3}(x^{2/3} - 1) + (x - 1) + o(1)\,,$$

$$n\log\frac{r}{R}\ =\ \frac{n}{3}\,\log x\ +\ \frac{1}{3}\left(\frac{n}{3}\right)^{1/3}(x^{2/3} - 1) + o(1)\,,$$

$$\frac{C_2(r)}{x\,C_2(R)}\ =\ \left(1 + o(1)\right),$$

where, in all the formulae, $o(1) \to 0$ uniformly for $x \in W$ and r is the unique real positive root of the equation

$$3\,r^3 + r = n.$$

From (4.48) it follows that

$$(4.58)\quad f_{n3}(x) = x^{n/3}\exp\left\{\left(\frac{n}{3}\right)^{1/3}(x^{2/3} - 1) + (x - 1)\right\}\left(1 + o(1)\right),$$

where $o(1) \to 0$ uniformly for $x \in W$.

Now, applying Theorem 4.2, we conclude that the distribution of the corresponding random variable ξ_n is asymptotically normal.

We note that

$$M_n = \frac{2}{3}\left(\frac{n}{3}\right)^{1/3} + 1 + o(1)\,, \qquad \sigma_n^2 = \left(\frac{2}{3}\right)^2\left(\frac{n}{3}\right)^{1/3}\left(1 + o(1)\right)$$

in the case under consideration. If m is arbitrary then the method described for finding an asymptotic representation for $f_{nm}(x)$ with explicit dependence on x only fails because it requires an asymptotic expansion for R, similar to (4.57), with knowledge of a large number of terms.

Finally, we consider the case where $s = 0$.

Theorem 4.4 *If $a_0 a_m > 0$ and*

(4.59) $$g(t) = a_0 + a_m t^m$$

then the distribution of the random variable $\eta_n = \xi_n - n/m$ is Poisson with parameter $\lambda = a_0$. If $a_0 = 0$ in (4.59) then the distribution of η_n is degenerate.

Proof We know that in this case

$$Q_{nm}(x) = \begin{cases} \dfrac{n!}{(n/m)!} \, (a_m x)^{n/m} \exp\{a_0 x\}, & m \mid n, \\[2mm] 0, & \text{otherwise}. \end{cases}$$

Hence it follows that, provided m divides n, the generating function of ξ_n is of the form

$$f_{nm}(x) = x^{n/m} \exp\{a_0 (x - 1)\}.$$

The generating function of η_n is equal to $\exp\{a_0 (x - 1)\}$, that is, it is the generating function of Poisson distribution with parameter $\lambda = a_0$. Finally, if $a_0 = 0$ and $m \mid n$ then $\mathbf{P}\{\xi_n = n/m\} = 1$, that is, the distribution of $\{\xi_n\}$ is concentrated at the point n/m.   □

CHAPTER 2

Combinatorial Properties of
Random Nonnegative Matrices

2.0 Introduction

Recent years have seen the publication of numerous papers which study
various combinatorial properties of nonnegative matrices. Certain characteristics of one or another class of matrix are usually considered that
create the necessary prerequisities for studying the corresponding probabilistic problems. The results of special interest are those concerned with
the asymptotic properties of the characteristics under investigation when
the sizes of matrices grow to infinity, and these are discussed in the current
chapter.

We give a short summary of the notions and facts which will be used
throughout the chapter. We denote by $2^{\mathbf{X}}$ the aggregate of subsets of the
set $\mathbf{X}$. A tuple of elements $(x_1, x_2, \ldots, x_n)$ is called a *transversal* of a
family of subsets $X_1, X_2, \ldots, X_n \in 2^{\mathbf{X}}$ if $x_i \in X_i$ and $x_i \neq x_j$, $i \neq j$,
$1 \leq i, j \leq n$. For a transversal of the family $X_1, X_2, \ldots, X_n$ to exist it is
necessary and sufficient that the inequalities

$$(0.1) \qquad \left| X_{i_1} \cup X_{i_2} \cup \ldots \cup X_{i_k} \right| \geq k$$

hold for $1 \leq k \leq n$, $1 \leq i_1 < i_2 < \cdots < i_k \leq n$, where $|\mathbf{Y}|$ means the
number of elements of the set $\mathbf{Y}$. Conditions (0.1) are usually termed Hall's
conditions.

A matrix $\mathrm{A} = \|a_{ij}\|$, $i = 1, 2, \ldots, n$, $j = 1, 2, \ldots, m$, is called the
incidence matrix of a family of subsets $X_1, X_2, \ldots, X_n$ of a set $\mathbf{X} =
\{x_1, x_2, \ldots, x_m\}$ if

$$a_{ij} = \begin{cases} 1 & \text{if } x_j \in X_i, \\ 0 & \text{if } x_j \notin X_i. \end{cases}$$

The number of transversals of the family $X_1, X_2, \ldots, X_n$ is equal to the permanent of the incidence matrix A which is defined (for $n \le m$) by the equality

$$\operatorname{Per} A = \sum_{j_1, \ldots, j_n} a_{1, j_1}\, a_{2, j_2} \cdots a_{n, j_n}\,,$$

where summation is taken over all ordered combinations of size n constituted by the numbers $1, 2, \ldots, m$.

If A is a matrix whose elements are all nonnegative, then, according to the König–Frobenius theorem, $\operatorname{Per} A = 0$ if and only if A contains a zero r by s submatrix with $r + s = m + 1$.

If $A = \| a_{ij} \|$, $i, j = 1, 2, \ldots, n$, is a double stochastic matrix, that is,

$$\sum_{j=1}^{n} a_{ij} = \sum_{i=1}^{n} a_{ij} = 1\,, \qquad a_{ij} \ge 0\,,$$

then $0 < \operatorname{Per} A \le 1$, where the upper bound is attained if and only if A is a permutation matrix. Recall that a square $(0, 1)$-matrix A is called a permutation matrix if each of its rows and columns contains exactly one element equal to 1. The rows (as well as the columns) of any permutation matrix constitute a set of noncollinear unit vectors. For this reason the elements of a permutation matrix equal to 1 are called noncollinear.

Let Ω_n be the set of n by n double stochastic matrices. In 1926 van der Waerden [126] formulated his celebrated conjecture: if $A \in \Omega_n$ and $A \ne J_n = \| n^{-1} \|$, then $\operatorname{Per} A > \operatorname{Per} J_n$. This conjecture was not proved until 1980, by Egorychev [29], who used some results of A.D. Aleksandrov [1] and London [70]. Simultaneously an independent proof of the van der Waerden conjecture was obtained by Falikman [41] but he did not show that J_n is a unique minimizing matrix.

Harper [52] gives the following probabilistic interpretation for the permanent of a double stochastic matrix $A = \| a_{ij} \|$, $i, j = 1, 2, \ldots, n$. Consider n boxes labeled $1, 2, \ldots, n$, each of which initially contains exactly one ball. Then all the balls are moved simultaneously and at random into other boxes with the probability that the ball contained in the ith box moves into the jth box is a_{ij}. In this case $0 < \operatorname{Per} A \le 1$ is the probability of the event that after such random movement each box will contain exactly one ball. Since the validity of the van der Waerden conjecture is confirmed, the minimal value of the probability is

$$\operatorname{Per} J_n = n! \cdot n^{-n}.$$

On the symmetric group $\mathbb{S}_n$ of order n we introduce a metric ρ by setting

$$\rho(s, s') := \big| \{\, i\colon s(i) \ne s'(i); \ s, s' \in \mathbb{S}_n, \ 1 \le i \le n \,\} \big|.$$

It is not difficult to check that ρ is indeed a metric. We say that permutations s and s' are *disjoint in k elements*, $0 \leq k \leq n$, if $\rho(s, s') = k$. In the case where $\rho(s, s') = n$ the permutations s and s' are said to be *disjoint*.

With each permutation s of order n one can associate, in a natural way, a unique permutation matrix Π of order n. If permutations s and s' are disjoint then the corresponding permutation matrices Π and Π' satisfy the following condition: if an element of the matrix Π is equal to 1 then the corresponding element of the matrix Π' is equal to 0.

Let $s_1, s_2, \ldots, s_k$ be a set of permutations of order n and let Π_1, $\Pi_2, \ldots, \Pi_k$ be the set of corresponding permutation matrices. A permutation s is called disjoint with the set of permutations $s_1, s_2, \ldots, s_k$ if it is disjoint with each permutation of this set. Denote by $N_n(s_1, s_2, \ldots, s_k)$ the number of permutations of order n disjoint with the set of permutations $s_1, s_2, \ldots, s_k$. Put $\Pi_r := \| \pi_{ij}^{(r)} \|$, $1 \leq r \leq k$, and consider a $(0,1)$-matrix $A = \| a_{ij} \|$, $i, j = 1, 2, \ldots, n$, with elements

$$
a_{ij} = \begin{cases} 1 & \text{if } \pi_{ij}^{(r)} = 0 \quad \text{for all} \quad r = 1, 2, \ldots, k, \\ 0 & \text{otherwise}. \end{cases}
$$

The equality

$$
N_n(s_1, s_2, \ldots, s_n) = \text{Per } A
$$

explains why permanents play an important role in solving enumeration problems concerning disjoint permutations [94, 111].

2.1 Nonnegative integer-valued matrices

We focus our attention on the class of n by m matrices with integer nonnegative elements whose sums are equal to N. The number of such matrices coincides with the number of different allocations of N identical objects into nm cells and, therefore, is equal to $\binom{nm+N-1}{N}$. We consider the uniform probability distribution on the set of matrices and denote by $\xi(N; n, m)$ the number of zero lines (rows and columns) in a randomly chosen matrix.

We number the lines of the matrix (rows and columns) in a certain way and consider the aggregate of events $E_1, E_2, \ldots, E_{n+m}$, where E_j is the event that the jth line of a randomly chosen matrix consists of zero elements only. Using the known expression for the kth binomial moment of $\xi(N; n, m)$,

$$
B_k(N; n, m) = \sum_{1 \leq j_1 < \cdots < j_k \leq n+m} \mathbf{P}\big(E_{j_1} E_{j_2} \ldots E_{j_k} \big),
$$

where $\mathbf{P}\left(E_{j_1}E_{j_2}\cdots E_{j_k}\right)$ is the probability of the simultaneous occurrence of events $E_{j_1}, E_{j_2}, \ldots, E_{j_k}$, it is not difficult to check that

$$(1.1) \quad B_k(N; n, m) = \binom{nm + N - 1}{N}^{-1}$$

$$\times \sum_{i=0}^{k} \binom{n}{t}\binom{m}{k-i}\binom{(n-i)(m-k+i)+N-1}{N}$$

for $k = 0, 1, \ldots, n + m$.

Now by the inclusion–exclusion method we obtain the exact distribution of $\xi(N; n, m)$:

$$\mathbf{P}\{\xi(N; n, m) = r\} = \sum_{k=r}^{n+m} (-1)^{k-r}\binom{k}{r} B_k(N; n, m).$$

Theorem 1.1 *Let* $n = [\alpha m], 0 < \alpha \le 1$, *and* $N m^{-1} - \log m \to \gamma \in (-\infty, \infty)$ *as* $m \to \infty$. *Then*

$$\lim_{m \to \infty} \mathbf{P}\{\xi(N; n, m) = r\} = \frac{\lambda^r}{r!}\, e^{-\lambda}, \qquad r = 0, 1, \ldots,$$

where $\lambda = 2\, e^{-\gamma}$ *if* $\alpha = 1$ *and* $\lambda = e^{-\gamma}$ *if* $0 < \alpha < 1$.

Proof Using the estimate

$$\binom{(n-i)(m-k+i)+N-1}{N}\binom{nm+N-1}{N}^{-1}$$

$$= \exp\left\{-N\left(\frac{i}{n} + \frac{k-i}{m}\right) + o(1)\right\},$$

one can deduce from (1.1) an asymptotic representation for the kth binomial moment of $\xi(N; n, m)$:

$$B_k(N; n, m) = \frac{1}{k!}\left(\exp\left\{\log m - \frac{N}{m}\right\} + \exp\left\{\log n - \frac{N}{n}\right\}\right)^k (1 + o(1)).$$

Hence it follows that

$$(1.2) \qquad B_k(N; n, m) = \frac{1}{k!}\left(e^{-\gamma} + \frac{\alpha\, e^{-\gamma/\alpha}}{m^{1/\alpha - 1}}\right)^k (1 + o(1)).$$

For any k the limit of the right-hand side of (1.2) is, as $m \to \infty$, $(2\, e^{-\gamma})^k / k!$ if $\alpha = 1$ and $e^{-k\gamma}/k!$ if $0 < \alpha < 1$. Hence, Theorem 1.1 follows. $\qquad\square$

We now consider a two-dimensional random variable (η_1, η_2), where η_1 is the number of zero rows and η_2 is the number of zero columns in a matrix chosen at random from the class of matrices under consideration. We have

(1.3) $\mathbf{P}\{\eta_1 = \mu, \eta_2 = \nu\}$

$$= \binom{n}{\mu}\binom{m}{\nu}\binom{(n-\mu)(m-\nu) + N - 1}{N}$$

$$\times \binom{nm + N - 1}{N}^{-1} \mathbf{P}\{\xi(N; n - \mu, m - \nu) = 0\}.$$

Corollary *Let the conditions of Theorem 1.1 hold. If $\alpha = 1$ then the distribution of the random variable (η_1, η_2) converges, as $m \to \infty$, to a two-dimensional Poisson distribution with independent components and parameters $(e^{-\gamma}, e^{-\gamma})$, that is,*

$$\lim_{m \to \infty} \mathbf{P}\{\eta_1 = \mu, \eta_2 = \nu\} = \frac{e^{-\gamma(\mu+\nu)}}{\mu! \, \nu!} \exp\{-2 e^{-\gamma}\}, \quad \mu, \nu = 0, 1, \dots.$$

If $0 < \alpha < 1$ then the distribution of η_1 is asymptotically degenerate and the distribution of η_2 converges, as $m \to \infty$, to a Poisson distribution with parameter $e^{-\gamma}$.

The corollary follows immediately from the asymptotic analysis of formula (1.3) by applying Theorem 1.1.

Below we consider a number of problems which may be characterized as follows. From the class of all n by m matrices with integral nonnegative elements we extract a subclass of matrices satisfying a set of properties Λ such that, for any matrix from this subclass, each of the properties preserves under deletion of a zero line from the matrix. We denote by $A(n, m; \Lambda)$ the number of matrices in the given subclass and by $A_r(n, m; \Lambda)$ the number of such submatrices with r zero lines. By the inclusion–exclusion method we obtain

(1.4) $$A_r(n, m; \Lambda) = \sum_{k=r}^{n+m} (-1)^{k-r} \binom{k}{r} S_k(n, m; \Lambda),$$

where

(1.5) $$S_k(n, m; \Lambda) = \sum_{i=0}^{k} \binom{n}{i}\binom{m}{k-i} A(n - i, m - k + i; \Lambda).$$

We now consider the uniform distribution on the subclass of matrices and denote by $\xi_{nm}(\Lambda)$ the number of zero lines in a randomly chosen matrix. Clearly,

$$\mathbf{P}\{\xi_{nm}(\Lambda) = r\} = A_r(n, m; \Lambda) / A(n, m; \Lambda),$$

and the quantity

$$B_k(n, m; \Lambda) = S_k(n, m; \Lambda) \, / \, A(n, m; \Lambda)$$

is the kth binomial moment of $\xi_{nm}(\Lambda)$, $k = 0, 1, \dots$. As a particular case, we refer to the class of n by m matrices the elements of which are independent and where each of them takes the values $0, 1, \dots, s-1$ with probability s^{-1}. If $\xi^{(s)}(n, m)$ is the number of zero lines in a matrix chosen at random and equiprobably from the described class then formulae (1.4) and (1.5) give

$$(1.6) \qquad \mathbf{P}\{\xi^{(s)}(n, m) = r\} = \sum_{k=r}^{n+m} (-1)^{k-r} \binom{k}{r} B_k^{(s)}(n, m),$$

where

$$(1.7) \quad B_k^{(s)}(n, m) = \sum_{j=0}^{k} \binom{n}{j} \binom{m}{k-j} s^{-j(m-k+j)-(k-j)n}, \quad k = 0, 1, \dots,$$

are binomial moments of $\xi^{(s)}(n, m)$.

Theorem 1.2 *Let $s > 1$, $\lambda > 0$ and $m = [\lambda s^n]$. Then the distribution of the random variable $\xi^{(s)}(n, m)$ converges, as $m \to \infty$, to a Poisson distribution with parameter λ.*

Proof Under the conditions of the theorem only a single summand in (1.7), where $j = 0$, does not vanish as $n \to \infty$. Since the summand tends to $\lambda^k/k!$ as $n \to \infty$, Theorem 1.1 follows. □

In the class of n by m matrices whose elements take the values $0, 1, \dots,$ $s - 1$ we consider the subclass of matrices having no zero rows. The total number of such matrices is $(s^m - 1)^n$. The number of such matrices with r zero columns is equal to

$$C_{nm}^{(s)}(r) = \sum_{k=r}^{m} (-1)^{k-r} \binom{k}{r} \binom{m}{k} \left(s^{m-k} - 1\right)^n.$$

Setting $r = 0$ in the preceding relation we find the number of matrices with no zero lines:

$$C_{nm}^{(s)} = \sum_{k=0}^{m} (-1)^k \binom{m}{k} \left(s^{m-k} - 1\right)^n.$$

By interchanging n and m we obtain

$$(1.8) \qquad C_{nm}^{(s)} = \sum_{k=0}^{n} (-1)^k \binom{n}{k} \left(s^{n-k} - 1\right)^m.$$

Let $\eta^{(s)}(n, m)$ be the number of zero columns in a matrix chosen at random and equiprobably in the class of matrices under consideration. Then

$$(1.9) \qquad \mathbf{P}\{\eta^{(s)}(n, m) = r\} = \sum_{k=r}^{m} (-1)^{k-r} \binom{k}{r} B_k(n, m, s),$$

where

$$(1.10) \qquad B_k(n, m, s) = \binom{m}{k} \left(\frac{s^{m-k} - 1}{s^m - 1} \right)^n$$

are binomial moments of $\eta^{(s)}(n, m)$. Formula (1.9) can be rewritten in the following form

$$(1.11) \quad \mathbf{P}\{\eta^{(s)}(n, m) = r\}$$

$$= \binom{m}{r} \sum_{j=0}^{m-r} (-1)^j \binom{m-r}{j} \left(\frac{s^{m-r-j} - 1}{s^m - 1} \right)^n.$$

From (1.11) it follows that, as $n \leq m$ and $m \to \infty$,

$$(1.12) \qquad \mathbf{P}\{\eta^{(s)}(n, m) = r\} = \mathbf{P}\{\widetilde{\eta}^{(s)}(n, m) = r\} (1 + o(1)),$$

where $\widetilde{\eta}^{(s)}(n, m)$ is a random variable having binomial distribution corresponding to the number of successes in m independent trials with probability of success s^{-n} in each trial. Formula (1.12), in combination with the known properties of binomial distribution, gives the following statement.

Theorem 1.3 *Let $n \leq m$ and $n \to \infty$. Then the following three assertions are valid:*

(a) *If $ms^{-n} \to \infty$ then the distribution of the random variable*

$$\left(\eta^{(s)}(n, m) - M \right) \sigma^{-1},$$

> *where*
> $$M := m\, s^{-n}, \qquad\qquad \sigma^2 := m\, s^{-n} \left(1 - s^{-n}\right),$$
> *is asymptotically normal with parameters $(0, 1)$.*

(b) *If $m\, s^{-n} \to \lambda \in (0, \infty)$ then the limiting distribution of $\eta^{(s)}(n, m)$ is Poisson with parameter λ.*

(c) *If $m\, s^{-n} \to 0$ then the distribution of $\eta^{(s)}(n, m)$ is asymptotically degenerate at the point 0.*

In conclusion we consider the class of $(0,1)$-matrices of order n by m, each of which has exactly N unit elements. If $\xi_{nm}(N)$ is the number of zero lines in a matrix chosen at random and equiprobably from this class then

$$\mathbf{P}\{\xi_{nm}(N) = r\} = \sum_{k=r}^{n+m} (-1)^{k-r} \binom{k}{r} B_k(n,m;N),$$

where

$$(1.13) \quad B_k(n,m;N) = \binom{nm}{N}^{-1} \sum_{i=0}^{k} \binom{n}{i}\binom{m}{k-i}\binom{(n-i)(m-k+i)}{N}.$$

Theorem 1.4 *Let* $n = [\alpha m]$, $0 < \alpha \leq 1$, *and* $N m^{-1} - \log m \to \gamma$ *as* $m \to \infty$. *Then the distribution of the random variable* $\xi_{nm}(N)$ *converges, as* $m \to \infty$, *to a Poisson distribution with parameter* λ, *where* $\lambda = 2 e^{-\gamma}$ *if* $\alpha = 1$ *and* $\lambda = e^{-\gamma}$ *if* $0 < \alpha < 1$.

Proof The proof of the theorem is parallel to that of Theorem 1.12 and is based on the asymptotic estimates of binomial moments of $\xi_{nm}(N)$ as $m \to \infty$. $\qquad\square$

2.2 Permanents of random (0,1)-matrices

We consider the set $\mathfrak{U}_{nm}$ of all $(0,1)$-matrices $A = \| a_{ij} \|$, $i = 1, 2, \ldots, n$, $j = 1, 2, \ldots, m$, $n \leq m$, and denote by $\mathfrak{U}_{nm}^1$ the set of matrices $A \in \mathfrak{U}_{nm}$ having either one zero row or at least $m - n + 1$ zero columns. We specify on $\mathfrak{U}_{nm}$ the uniform probability distribution and study the probabilities of the events

$$H_{nm}^1 := \{ A \in \mathfrak{U}_{nm}^1 \}, \qquad E_{nm} := \{ \operatorname{Per} A = 0, A \in \mathfrak{U}_{nm} \}.$$

It is easily seen that

$$(2.1) \qquad \mathbf{P}\big(H_{nm}^1 \mid E_{nm}\big) = 1.$$

In fact, a more general statement is valid. It appears that if $n \leq m$ and $m \to \infty$, then the occurrence of the event E_{nm} implies that of the event H_{nm}^1 with probability close to 1. We formulate the statement as a theorem.

Theorem 2.1 *If A is a matrix chosen at random and equiprobably from $\mathfrak{U}_{nm}$ then*

$$(2.2) \qquad \lim_{m \to \infty} \mathbf{P}\big(H_{nm}^1 \mid E_{nm}\big) = 1$$

uniformly for $n \leq m$.

Proof Denoting by $\bar{H}_{nm}^1$ the event complementary to H_{nm}^1 we obtain

$$(2.3) \qquad \mathbf{P}(H_{nm}^1 \mid E_{nm}) = 1 - \frac{\mathbf{P}(\bar{H}_{nm}^1 \cap E_{nm})}{\mathbf{P}(E_{nm})}.$$

Evidently, $\mathbf{P}(\bar{H}_{nm}^1 \cap E_{nm})$ is the probability that a random matrix $A \in \mathfrak{U}_{nm}$ satisfies the condition Per $A = 0$ and has at most $m - n$ zero columns and no zero rows.

Let B_{rs} be the event that a random matrix $A \in \mathfrak{U}_{nm}$ contains a zero r by s submatrix. By the König–Frobenius theorem each matrix $A \in \bar{H}_{nm}^1 \cap E_{nm}$ has a zero r by s submatrix with $r + s = m + 1$, $2 \leq r \leq n - 1$, and, therefore, the inclusion

$$\bar{H}_{nm}^1 \cap E_{nm} \subseteq \bigcup_{r=2}^{n-1} B_{r,\, m-r+1}$$

occurs. Applying the obvious estimate

$$\mathbf{P}(B_{rs}) \leq \binom{n}{r}\binom{m}{s} 2^{-rs},$$

we have

$$\mathbf{P}(\bar{H}_{nm}^1 \cap E_{nm}) \leq \sum_{r=2}^{m-1} \binom{m}{r}\binom{m}{r-1} 2^{-r(m-r+1)}.$$

Hence, it follows that

$$(2.4) \quad \mathbf{P}(\bar{H}_{nm}^1 \cap E_{nm}) \leq m^3\, 2^{-2(m-1)}$$
$$+ \frac{1}{6}\, m^5\, 2^{-3(m-2)} + \binom{2m}{m+1} 2^{-4(m-3)}.$$

We now search for a lower bound of probability $\mathbf{P}(E_{nm})$. Let L_{nm} be the event that a random matrix $A \in \mathfrak{U}_{nm}$ has no zero rows. Obviously,

$$(2.5) \qquad \mathbf{P}(\bar{H}_{nm}^1) \leq \mathbf{P}(L_{nm}) = (1 - 2^{-m})^n$$

and, therefore,

$$\mathbf{P}(E_{nm}) \geq \mathbf{P}(H_{nm}^1) \geq 1 - (1 - 2^{-m})^n.$$

Using induction on n it is not difficult to show that

$$(1 - 2^{-m})^n \leq 1 - n\,2^{-m} + \binom{n}{2} 2^{-2m}.$$

Combining (2.4) and (2.5) we obtain, as $m \to \infty$, the estimates

$$(2.6) \qquad \mathbf{P}\left(\bar{H}_{nm}^{1} \cap E_{nm}\right) = O\left(\frac{m^3}{2^{2m}}\right),$$

$$(2.7) \qquad \frac{\mathbf{P}\left(\bar{H}_{nm}^{1} \cap E_{nm}\right)}{\mathbf{P}\left(E_{nm}\right)} = O\left(\frac{m^3}{2^{m}}\right),$$

which are uniform for $2 \le n \le m$. Now Theorem 2.1 follows from (2.3) and (2.7). $\qquad \square$

Observe that estimate (2.7) can be improved if $m = n$, namely, in this case

$$(2.8) \qquad \frac{\mathbf{P}\left(\bar{H}_{nn}^{1} \cap E_{nn}\right)}{\mathbf{P}\left(E_{nm}\right)} = O\left(\frac{n^2}{2^{n}}\right).$$

One can also obtain a number of estimates for the probabilities of certain events related to a random matrix $A \in \mathfrak{U}_{nm}$. To be more specific, denote by $\mathfrak{U}_{nm}^{0}$ and $\mathfrak{U}_{nm}^{00}$, respectively, the sets of matrices $A \in \mathfrak{U}_{nm}$ with no zero rows and no zero lines and put

$$(2.9) \qquad C_{nm} := |\mathfrak{U}_{nm}^{00}|,$$

$$(2.10) \qquad N_{nm} := \left| \left\{ A \colon \operatorname{Per} A = 0, \ A \in \mathfrak{U}_{nm}^{00} \right\} \right|.$$

We consider the uniform distribution on the set $\mathfrak{U}_{nm}^{0}$ and denote by $H_{nm}^{(j)}$ the event that a random matrix $A \in \mathfrak{U}_{nm}^{0}$ has exactly j zero columns. Setting

$$H_{nm} := \bigcup_{j=m-n+1}^{m-1} H_{nm}^{(j)},$$

we conclude that the probability $P_{nm}^{0}\{\operatorname{Per} A = 0\}$ of the event that the permanent of a matrix A chosen at random from the set $\mathfrak{U}_{nm}^{0}$ equal 0 is given by the formula

$$(2.11) \qquad P_{nm}^{0}\{\operatorname{Per} A = 0\} = \mathbf{P}\left(H_{nm}\right) + \sum_{j=0}^{m-n} \mathbf{P}\left(H_{nm}^{(j)} \cap \{\operatorname{Per} A = 0\}\right).$$

Using notations (2.9) and (2.10) one can write

$$(2.12) \qquad \mathbf{P}\left(H_{nm}^{(j)}\right) = \binom{m}{j} C_{n,\,m-j}\,(2^m - 1)^{-n}, \quad j = 0, 1, \ldots, m-1,$$

$$(2.13) \quad \mathbf{P}\left(H_{nm}^{(j)} \cap \{\operatorname{Per} A = 0\}\right)$$
$$= \binom{m}{j} N_{n,\,m-j}\,(2^m - 1)^{-n}, \ j = 0, 1, \ldots, m-1.$$

Now combining (2.11)–(2.13) we find that

$$(2.14) \quad P_{nm}^0\{\,\text{Per A} > 0\,\} = (2^m - 1)^{-n} \sum_{j=0}^{m-n} \binom{m}{j}\left(C_{n,\,m-j} - N_{n,\,m-j}\right).$$

Setting

$$P_n^0\{\,\text{Per A} > 0\,\} := P_{nn}^0\{\,\text{Per A} > 0\,\}$$

and

$$C_n := C_{nn}, \qquad\qquad N_n := N_{nn},$$

we derive the following from (2.14):

$$(2.15) \qquad P_n^0\{\,\text{Per A} > 0\,\} = (C_n - N_n)\,(2^n - 1)^{-n}.$$

From (1.8), taking $s = 2$, we know that

$$(2.16) \qquad C_{nm} = \sum_{k=0}^{m}(-1)^k \binom{m}{k}\left(2^{m-k} - 1\right)^n$$

and, by Bonferroni's inequalities,

$$(2.17) \qquad \left(2^m - 1\right) - m\left(2^{m-1} - 1\right)^n \le C_{nm} \le \left(2^m - 1\right)^n.$$

By the König–Frobenius theorem one can obtain the estimate

$$(2.18) \qquad N_{nm} \le \sum_{r=2}^{n-1} \binom{n}{r}\binom{m}{m-r+1} 2^{\,nm - r\,(m-r+1)}.$$

Equalities (2.17) and (2.18) allow us to derive the simpler estimates

$$(2.19) \qquad C_{nm} \ge \left(2^m - 1\right)^n\left(1 - m\,2^{-n}\right),$$

$$(2.20) \qquad N_{nm} \le 3 \cdot 2^{nm} / 2^{m-4},$$

which can serve as a basis for the proof of the following theorem.

Theorem 2.2 *For* $n \le m \le 2^n$

$$(2.21) \quad P_{nm}^0\{\,\text{Per A} > 0\,\} \ge 1 - \frac{m}{2^n} - \frac{3}{2^{n-4}}\left(1 - \frac{n}{2^m}\right)^{-1} \sum_{j=0}^{m-n}\binom{m}{j}\frac{1}{2^{nj}}.$$

Proof Using (2.19) we find that

$$(2.22) \qquad \frac{1}{\left(2^m - 1\right)^n} \sum_{j=0}^{m-n}\binom{m}{j} C_{n,\,m-j} \ge \left(1 - \frac{m}{2^n}\right)$$

$$\times\left[1 + \left(\frac{1 - 2^{-n}}{1 - 2^{-m}}\right)^n \sum_{j=1}^{m-n}\binom{m}{j}\frac{1}{2^{nj}}\right].$$

According to (2.20),

$$(2.23) \qquad \frac{1}{(2^m - 1)^n} \sum_{j=0}^{m-n} \binom{m}{j} N_{n, m-j}$$

$$\leq \frac{3}{2^{n-4}} \left(1 - \frac{n}{2^m}\right)^{-1} \sum_{j=0}^{m-n} \binom{m}{j} \frac{1}{2^{nj}}.$$

Now estimate (2.21) follows from (2.14) and inequalities (2.22) and (2.23). $\qquad\square$

Corollary *The following relations are valid:*

$$(2.24) \qquad P_n^0\{\operatorname{Per} A > 0\} \geq 1 - \frac{n}{2^n} - \frac{3}{2^{n-4}} \left(1 - \frac{n}{2^n}\right)^{-1}$$

and

$$(2.25) \qquad \lim_{n\to\infty} P_n^0\{\operatorname{Per} A > 0\} = 1.$$

Equality (2.25) allows us to prove one more statement.

Theorem 2.3 *If $n \leq m$ then*

$$(2.26) \qquad \lim_{n\to\infty} P_{nm}^0\{\operatorname{Per} A > 0\} = 1.$$

Proof Let $A \in \mathfrak{U}_{nm}^0$ and let A' be an n by n submatrix of A. Obviously, as $n \to \infty$,

$$(2.27) \qquad P_{nm}^0\{A' \in \mathfrak{U}_{nn}^0\} = \frac{(2^n - 1)^n \, 2^{(m-n)\,n}}{(2^m - 1)^n} \longrightarrow 1.$$

The uniform distribution on the set of matrices $\mathfrak{U}_{nm}^0$ induces the uniform distribution on the set of matrices $A' \in \mathfrak{U}_{nn}^0$. Thus, as $n \to \infty$, equality (2.25) gives

$$(2.28) \qquad P_{nm}^0(\{\operatorname{Per} A' > 0\} \mid \{A' \in \mathfrak{U}_{nn}^0\}) \longrightarrow 1.$$

Relations (2.27) and (2.28) imply

$$(2.29) \qquad P_{nm}^0(\{\operatorname{Per} A' > 0\} \cap \{A' \in \mathfrak{U}_{nn}^0\}) \longrightarrow 1, \qquad n \to \infty.$$

Now (2.26) follows from the inequality

$$P_{nm}^0\{\operatorname{Per} A > 0\} \geq P_{nm}^0(\{\operatorname{Per} A' > 0\} \cap \{A' \in \mathfrak{U}_{nn}^0\})$$

and relation (2.26). $\qquad\square$

2.3 Transversals of random sets

2.3.1 Limit theorems

We consider the uniform distribution on the set $2^{\mathbf{X}}$, the set of subsets of an m-element set $\mathbf{X}$. Let us take randomly and independently nonempty sets $X_1, X_2, \ldots, X_n \in 2^{\mathbf{X}}$.

Theorem 3.1 *Let $n \leq m$ and $n \to \infty$. Then with probability tending to 1 one can find distinct elements $x_1, x_2, \ldots, x_n \in \mathbf{X}$ constituting a transversal of the sets $X_1, X_2, \ldots, X_n$.*

Proof Let A be the incidence matrix of the sets $X_1, X_2, \ldots, X_n$ and the elements of the set $\mathbf{X} = \{x_1, x_2, \ldots, x_m\}$. Then the desired transversal exists if and only if $\{\operatorname{Per} A > 0\} > 0$. Clearly, A has no zero rows and, therefore, the probability of positiveness of its permanent under random choice of $X_1, X_2, \ldots, X_n$ is $P_{nm}^0\{\operatorname{Per} A > 0\}$, where we have used the notation of the preceding section. According to (2.26) this probability tends to 1 as $n \to \infty$. □

We now refine Theorem 3.1 slightly. With this aim, we consider a random matrix $A \in \mathfrak{U}_{nm}$ and denote by $\widetilde{P}_{nm}$ the probability of the event that each n by n submatrix A' of A has no zero rows. The probability $\widetilde{P}_{nm}$ is calculated by means of the formula

$$(3.1) \qquad \widetilde{P}_{nm} = \left(1 - \frac{\gamma_{nm}}{n} \right)^n,$$

where

$$\gamma_{nm} = \frac{n}{2^m - 1} \sum_{k=n}^{m-1} \binom{m}{k}.$$

Put $\alpha = \left(n - \tfrac{1}{2} m \right) \left(\tfrac{1}{4} m \right)^{-1/2}$.

Theorem 3.2 *Let n and m tend to infinity in such a way that*

$$(3.2) \qquad \frac{n}{\alpha \sqrt{2\pi}} \exp\left\{ -\tfrac{1}{2} \alpha^2 \right\} \longrightarrow \gamma \in [0, \infty].$$

Then

$$(3.3) \qquad \lim \widetilde{P}_{nm} = \begin{cases} 0 & \text{if } \gamma = \infty, \\ e^{-\gamma} & \text{if } 0 < \gamma < \infty, \\ 1 & \text{if } \gamma = 0. \end{cases}$$

Remark Direct calculations show that if $\gamma \in (0, \infty)$ in (3.2), then

$$n = \frac{m}{2} + \sqrt{\frac{m \log m}{2}} - \frac{\sqrt{m} \, \log \log m}{4 \sqrt{2 \log m}}$$

$$- frac\log (4 \, \gamma \sqrt{\pi}) 2 \sqrt{2} \, \sqrt{\frac{m}{2 \log m}} + o\left(\frac{m}{\log m}\right)$$

as $m \to \infty$.

Proof of Theorem 3.2 It follows from the definition of γ_{nm} that, as $m \to \infty$,

$$\gamma_{nm} \sim n \, \mathbf{P} \{ m > \xi_m \geq n \} \sim n \sum_{k=n}^{m-1} \binom{m}{k} 2^{-m}$$

$$= n \left(\mathbf{P} \{ \xi_m \geq n \} - 2^{-m} \right) = n \, \mathbf{P} \{ \xi_m > n \} + o(1),$$

where ξ_m is a random variable having binomial distribution with parameters $(m, \frac{1}{2})$. Set

$$\xi_m^* := \left(\xi_m - \tfrac{1}{2} m \right) \left(\tfrac{1}{4} m \right)^{-1/2}.$$

It is known (see, for example, [42, Vol. 1, Chapter VII, Section 6]) that

$$\mathbf{P} \{ \xi_m^* \geq x \} \sim \frac{1}{x \sqrt{2\pi}} \exp \left\{ -\tfrac{1}{2} x^2 \right\}, \qquad x \to \infty,$$

provided that $x = o(m^{1/6})$ as $m \to \infty$ and, obviously,

$$\frac{1}{x \sqrt{2\pi}} \exp \left\{ -\tfrac{1}{2} x^2 \right\} = o(m^{-1})$$

if $x > \sqrt{2 \log m}$. These estimates allow us to state: if condition (3.2) holds and $\alpha \to \infty$, then

$$\gamma_{nm} \sim n \, \mathbf{P} \{ \xi_m \geq n \} + o(1) = n \, \mathbf{P} \{ \xi_m \geq n \} + o(1) \to \gamma$$

as $m \to \infty$. If, under the same condition, α is bounded from above, then

$$\lim n \, \mathbf{P} \{ \xi_m^* > \alpha \} = \infty$$

by the central limit theorem applied to the binomial distribution in question. To complete the proof of Theorem 3.2 it suffices to note that

$$\widetilde{P}_{nm} \sim \exp \{ -\gamma_{nm} \} \quad \text{if} \quad \gamma_{nm} \to \gamma < \infty$$

and

$$\widetilde{P}_{nm} \to 0 \quad \text{if} \quad \gamma_{nm} \to \infty. \qquad \square$$

Theorem 3.3 *Let $n \leq m$, $n \to \infty$ and let condition (3.2) hold with $\gamma = 0$. Then, with probability tending to 1, any tuple of distinct elements $x_{j_1}, x_{j_2}, \ldots, x_{j_n} \in \mathbf{X}$, $|\mathbf{X}| = m$, is a transversal of nonempty sets $X_1, X_2, \ldots, X_n$ chosen at random and equiprobably from $2^{\mathbf{X}}$.*

Proof Let A be the incidence matrix of the sets $X_1, X_2, \ldots, X_n$ and the elements of the set $\mathbf{X}$ and let A' be the incidence matrix of $X_1, X_2, \ldots, X_n$ and a subset $\mathbf{X}' = \{ x_{j_1}, x_{j_2}, \ldots, x_{j_n} \} \subset \mathbf{X}$. To prove the theorem it suffices to establish that

$$(3.4) \qquad \lim_{n \to \infty} P_n\{\, \text{Per A}' > 0 \,\} = 1$$

for an arbitrary choice of $x_{j_1}, x_{j_2}, \ldots, x_{j_n}$. Because

$$\left(\text{Per A}' > 0\right) \supseteq \left(\text{A}' \in \mathfrak{U}_{nn}^0\right) \bigcap \left(\text{Per A}' > 0\right),$$

the following inequality is valid:

$$(3.5) \qquad P_n\{\, \text{Per A}' > 0 \,\} \geq \widetilde{P}_{nm} + P_n^0\{\, \text{Per A} > 0 \,\} - 1.$$

Relations (3.5), (2.25) and (3.3) yield (3.4).   □

We now give a description of the asymptotic conditions that provide the nonexistence of a transversal of random sets $X_1, X_2, \ldots, X_n \in \mathbf{X}$, $|\mathbf{X}| = m \geq n$. Denote by $\mathfrak{U}_{nm}^{(1)}$ (respectively, by $\mathfrak{U}_{nm}^{(2)}$) the set of matrices $\text{A} \in \mathfrak{U}_{nm}$ having exactly one zero column (row) and let $\mathfrak{U}_{nm}^* := \mathfrak{U}_{nm}^{(1)} \cup \mathfrak{U}_{nm}^{(2)}$. For nonempty sets $X_1, X_2, \ldots, X_n \in 2^{\mathbf{X}}$ consider two random events $\mathfrak{N}^{(1)}$ and $\mathfrak{N}^{(2)}$. The event $\mathfrak{N}^{(1)}$ occurs if exactly one set among $X_1, X_2, \ldots, X_n$ has no common elements with the set $\{ x_{j_1}, x_{j_2}, \ldots, x_{j_n} \}$. Clearly, $\mathfrak{N}_n^{(i)}$ occurs if and only if the corresponding incidence matrix A' belongs to the set $\mathfrak{U}_{nn}^{(i)}$, $i = 1, 2$.

Consider two random objects: a set $\{ x_{j_1}, x_{j_2}, \ldots, x_{j_n} \}$ consisting of distinct elements chosen sequentially, at random and equiprobably from the set $\mathbf{X} = \{ x_1, x_2, \ldots, x_n \}$, and a set of nonempty sets $X_1, X_2, \ldots, X_n$ chosen at random and equiprobably from the set $2^{\mathbf{X}}$. Let A be the incidence matrix of $X_1, X_2, \ldots, X_n$ and $\{ x_1, x_2, \ldots, x_n \}$ and let A' be the incidence matrix of $X_1, X_2, \ldots, X_n$ and $\{ x_{j_1}, x_{j_2}, \ldots, x_{j_n} \}$. Denote by $\mathfrak{M}_{nm}$ the event that $\{ x_{j_1}, x_{j_2}, \ldots, x_{j_n} \}$ is not a transversal of the sets $X_1, X_2, \ldots, X_n$. Clearly, $\mathfrak{M}_{nm} = \{ \text{Per A}' = 0 \}$.

The following theorem is valid.

Theorem 3.4 *The event $\mathfrak{M}_{nm}$ is asymptotically tantamount to the event $\mathfrak{N}^{(1)} \cup \mathfrak{N}^{(2)}$ as $n \to \infty$, $n < m$, that is,*

$$(3.6) \qquad \lim_{n \to \infty} \mathbf{P}\big(\mathfrak{N}^{(1)} \bigcup \mathfrak{N}^{(2)} \,\big|\, \mathfrak{M}_{nm}\big) = 1.$$

Proof To prove the theorem it suffices to show that

$$(3.7) \qquad \lim_{n \to \infty} \mathbf{P}\big(\{\, \text{A}' \in \mathfrak{U}_{nm}^* \,\} \,\big|\, \{\text{Per A}' = 0\}\big) = 1.$$

It is not difficult to establish that the probability in question satisfies the relation

(3.8) $$\mathbf{P}(\{\, A' \in \mathfrak{U}_{nn}^*\,\} \mid \{\, \mathrm{Per}\, A' = 0\,\}) = \frac{\mathbf{P}\{\, A' \in \mathfrak{U}_{nn}^*\,\}}{\mathbf{P}\{\, \mathrm{Per}\, A' = 0\,\}}\,.$$

Introduce the notations

$$Q_{nm}^* := |\mathfrak{U}_{nm}^*|, \qquad\qquad Q_{nn} := |\mathfrak{U}_{nm}^1|.$$

There are $n\,(2^m - 1)^{n-1}$ matrices in $\mathfrak{U}_{nm}^{(1)}$ and $m\,(2^n - 1)^{m-1}$ matrices in $\mathfrak{U}_{nm}^{(2)}$. Observe also that among the elements of the set of n by m $(0,1)$-matrices, exactly $mn\,C_{n-1,\,m-1}$ matrices have one zero row and one zero column. Thus,

(3.9) $$Q_{nm}^* = n\,(2^m - 1)^{n-1} + m\left(2^n - 1\right)^{m-1} - mn\,C_{n-1,\,m-1}$$

and, in particular,

(3.10) $$Q_{nn}^* = 2n\left(2^n - 1\right)^{n-1} - n^2\,C_{n-1,\,n-1}\,.$$

By the estimate $C_{n-1,\,n-1} \le (2^{n-1} - 1)^{n-1}$ we derive from (3.9) that

(3.11) $$\lim_{n\to\infty} \frac{Q_{nn}^*}{n\,2^{n^2-n+1}} = 1\,.$$

Retaining the notations of the preceding section, one can write

$$\mathbf{P}\left\{\, \mathrm{Per}\, A' = 0\,\right\} = \mathbf{P}\left\{\, A' \in \mathfrak{U}_{nn}^1\,\right\} + \mathbf{P}(\{\, A' \in \mathfrak{U}_{nm}^{00}\,\} \cap \{\, \mathrm{Per}\, A' = 0\,\})\,.$$

Recalling that

$$N_{nn} = \left|\,\{\, A:\ \mathrm{Per}\, A = 0,\ \ A \in \mathfrak{U}_{nm}^{00}\,\}\,\right|,$$

we find that

(3.12) $$\mathbf{P}(\{\, A' \in \mathfrak{U}_{nn}^*\,\} \mid \{\, \mathrm{Per}\, A' = 0\,\}) = Q_{nn}^*\left(Q_{nn} + N_{nn}\right)^{-1}.$$

The relation $C_n + Q_n = 2^{n^2}$ and representation (2.16) give

(3.13) $$\lim_{n\to\infty} \frac{Q_{nn}}{n\,2^{n^2-n+1}} = 1\,.$$

Finally, inequality (2.20) with $m = n$ yields

(3.14) $$\lim_{n\to\infty} \frac{N_{nn}}{n\,2^{n^2-n+1}} = 0\,.$$

Combining (3.11), (3.13) and (3.14) it is not difficult to deduce from (3.12) the desired result (3.7). $\qquad\Box$

We note that the condition $m < n$ is needed to provide the equally likely choice of the submatrix $A' \in \mathfrak{U}_{nm}$ that would be impossible if $m = n$ since $A \in \mathfrak{U}_{nm}^1$ by hypothesis. We now consider the case where $m = n$. Let $X_1, X_2, \dots, X_n$ be nonempty subsets of the set $\mathbf{X} = \{\, x_1, x_2, \dots, x_n\,\}$ chosen at random and equiprobably from the set $2^{\mathbf{X}}$. Denote by $\mathfrak{M}_n$ the event that $x_1, x_2, \dots, x_n$ is not a transversal of $X_1, X_2, \dots, X_n$, and by $\mathfrak{M}_n^{(1)}$ the event that a random matrix $A \in \mathfrak{U}_n^0$ has exactly one zero column.

Theorem 3.5 *The following equality is valid:*

(3.15)
$$\lim_{n\to\infty} \mathbf{P}\big(\mathfrak{N}^{(1)} \,|\, \mathfrak{M}_n\big) = 1.$$

Proof Obviously, it suffices to prove that

(3.16)
$$\lim_{n\to\infty} \mathbf{P}\big(\{A \in \mathfrak{U}_n^{(1)}\} \,|\, \{\,\mathrm{Per}\,A = 0\}\big) = 1,$$

where $\mathfrak{U}_n^{(1)}$ is the set of n by n $(0,1)$-matrices with no zero rows and exactly one zero column.

Introduce the notations

$$Q_n := \big|\,\{A\colon\, A \in \mathfrak{U}_{nm}^0 \setminus \mathfrak{U}_{nm}^{00}\}\,\big|, \qquad Q_n^* := \big|\,\{A\colon\, A \in \mathfrak{U}^{(1)}\}\,\big|.$$

Then the conditional probability in question can be written as

(3.17)
$$\lim_{n\to\infty} \mathbf{P}\big(\mathfrak{N}^{(1)} \,|\, \mathfrak{M}_n\big) = Q_n^* \big(Q_n + N_{nn}\big)^{-1}.$$

The equality $Q_n = (2^n - 1)^n - C_{nn}$ and formula (2.16) give

(3.18)
$$\lim_{n\to\infty} \frac{Q_n}{n\,2^{n^2-n}} = 1.$$

Eventually, using the equality $Q_n^* = n\,C_{n,n-1}$ and formula (2.16), we obtain

(3.19)
$$\lim_{n\to\infty} \frac{Q_n^*}{n\,2^{n^2-n}} = 1.$$

Now the validity of (3.16) follows from (3.14), (3.18), (3.19) and representation (3.17).   □

Note that the asymptotic behavior of the number of $(0,1)$-matrices with zero permanents was studied in [40].

In conclusion we would like to focus the reader's attention on the conditions of existence or nonexistence of a transversal for a family of nonempty subsets $X_1, X_2, \ldots, X_n \subseteq \mathbf{X}$, $|\mathbf{X}| = m$, for large $n \leq m$. The well-known Hall criterion of existence of a transversal necessitates verifying $2^n - 1$ inequalities. To apply the criterion to the incidence matrix A of the family in question we must check that for any k, $1 \leq k \leq n$, each k by m submatrix has at least k nonzero columns. On the other hand, Theorem 3.1 states that if $n \to \infty$, then, with probability close to 1, each family has a transversal. Moreover, under the additional condition (3.2) (with $\gamma = 0$) any set of elements $x_{j_1}, x_{j_2}, \ldots, x_{j_n}$ is a transversal of the family $X_1, X_2, \ldots, X_n$ with probability tending to 1. In searching for transversals a description of the most probable events under which the search fails is of special interest. It follows from Theorem 3.4 that if $m < n$ then, with probability close to 1, a random tuple of elements $x_{j_1}, x_{j_2}, \ldots, x_{j_n} \in \mathbf{X}$ is not a transversal of the family of nonempty subsets $X_1, X_2, \ldots, X_n \subseteq \mathbf{X}$ for one of two reasons: either one can find an element $x_j \in \mathbf{X}$ which does not belong to the union of $X_1, X_2, \ldots, X_n$, or one can indicate a subset X_i containing no common elements with $\{x_{j_1}, x_{j_2}, \ldots, x_{j_n}\}$. For $m = n$ with probability tending to 1, only one event is responsible for the nonexistence of a transversal: an element exists which does not belong to the union of $X_1, X_2, \ldots, X_n$.

2.4 Permanents of random (0,1)-matrices with a prescribed number of 1s

2.4.1 Conditions for the asymptotic equivalence of a random permanent to its mean value

We denote by $\mathfrak{U}(n, N)$ the set of all square $(0,1)$-matrices of order n with N elements equal to 1, $n \leq N \leq n^2$. We specify the uniform distribution on the set by assigning to every matrix $A \in \mathfrak{U}(n, N)$ the probability $\binom{n^2}{N}^{-1}$. Let us consider the random variable $\operatorname{Per} A$, where A is a matrix chosen at random from $\mathfrak{U}(n, N)$. We will look for its mean and its variance.

We arrange all the permutations of the symmetric group $\mathbb{S}_n$ in some way and write $\mathbb{S}_n = \{ s_1, s_2, \ldots, s_{n!} \}$. Let us consider the random variables $\xi_1, \xi_2, \ldots, \xi_{n!}$, where $\xi_i = \xi_i(A) = 1$ if the matrix $A \in \mathfrak{U}(n, N)$ has n noncollinear 1s generating the permutation matrix Π_i corresponding to the permutation s_i, and $\xi_i = 0$ otherwise. Obviously,

$$(4.1) \qquad \mathbf{P}\{ \xi_i = 1 \} = \binom{n^2 - n}{N - n} \Big/ \binom{n^2}{N}, \qquad i = 1, 2, \ldots, n!.$$

It is clear that any matrix $A \in \mathfrak{U}(n, N)$ admits the representation:

$$\operatorname{Per} A = \xi_1 + \xi_2 + \cdots + \xi_{n!}.$$

Hence the mean value of the permanent of a random matrix can be computed by the formula

$$(4.2) \qquad \mathbf{E}\left(\operatorname{Per} A\right) = n!\,(N)_n \big/ (n^2)_n.$$

Adopting similar arguments one can calculate the variance of the permanent of a random matrix. We have

$$\mathbf{E}\,\xi_i\xi_j = \mathbf{P}\{ \xi_i = 1, \xi_j = 1 \}$$

$$= \binom{n^2 - 2n + r}{N - 2n + r} \Big/ \binom{n^2}{N} = (N)_{2n-r} \big/ (n^2)_{2n-r},$$

where the quantity r is determined by the relation

$$\rho(s_i, s_j) = \big| \{ \nu \colon s_i(\nu) \neq s_j(\nu), \; \nu = 1, 2, \ldots, n \} \big| = n - r.$$

For a fixed permutation s_i the number A_{nr} of permutations s_j such that $\rho(s_i, s_j) = n - r$ is given by the formula

$$A_{nr} = \frac{n!}{r!} \sum_{k=0}^{n-r} \frac{(-1)^k}{k!}.$$

Therefore, the second moment of the permanent of a random matrix is

$$(4.3) \quad \mathbf{E} \left(\text{Per } A \right)^2 = \sum_{i,j} \mathbf{E} \, \xi_i \xi_j = \sum_{i=1}^{n!} \sum_{r=0}^{n} \sum_{j} \mathbf{E} \, \xi_i \xi_j \, 1_{\{\rho(s_i, s_j) = n - r\}}$$

$$= (n\,!)^2 \sum_{r=0}^{n} \frac{(N)_{2n-r}}{(n^2)_{2n-r}} \cdot \frac{1}{r!} \sum_{k=0}^{n-r} \frac{(-1)^k}{k!} \, .$$

Here 1_A is the indicator of the event A. The final expression for the variance of the permanent of a random matrix $A \in \mathfrak{U}(n, N)$ is

$$(4.4) \quad \text{Var} \left(\text{Per } A \right) = (n\,!)^2 \left[\sum_{r=0}^{n} \frac{(N)_{2n-r}}{(n^2)_{2n-r}} \cdot \frac{1}{r!} \sum_{k=0}^{n-r} \frac{(-1)^k}{k!} - \frac{(N)_n^2}{(n^2)_n^2} \right].$$

We now study the asymptotic behavior of the mean and variance of the permanent of a random $(0, 1)$-matrix $A \in \mathfrak{U}(n, N)$ as $n \to \infty$, assuming that the number of 1s in the matrix exceeds $n^{2/3+\varepsilon}$ for some $\varepsilon > 0$. Using simple asymptotic estimates it is not difficult to prove the following lemma.

Lemma 1 *If $m \to \infty$ and $\theta = o(m^{2/3})$ then*

$$\prod_{j=0}^{\theta-1} (m - j) = m^\theta \exp \left\{ - \frac{\theta (\theta - 1)}{2m} \right\} \left(1 + O \left(\frac{\theta^3}{m^2} \right) \right).$$

Lemma 2 *If $N > n^{2/3+\varepsilon}, \varepsilon > 0$ then, as $n \to \infty$, the mean value of the permanent of a random matrix $A \in \mathfrak{U}(n, N)$ admits the asymptotic representation*

$$(4.5) \quad \mathbf{E} \left(\text{Per } A \right) = n! \left(\frac{N}{n^2} \right)^n \exp \left\{ - \frac{1}{2} \left(\frac{n^2}{N} - 1 \right) \right\} \left(1 + O \left(\frac{1}{n^{2\varepsilon}} \right) \right).$$

Proof Applying Lemma 1 with $\theta = n$, $m = N$ and $m = n^2$ we obtain

$$(4.6) \qquad (N)_n = N^n \exp \left\{ - \frac{n (n - 1)}{2N} \right\} \left(1 + O \left(\frac{n^3}{N^2} \right) \right),$$

$$(4.7) \qquad (n^2)_n = n^{2n} \exp \left\{ - \frac{1}{2} \right\} \left(1 + O \left(\frac{1}{n} \right) \right).$$

Substituting (4.6) and (4.7) into formula (4.2) we prove (4.5).   □

Lemma 3   *Under the conditions of Lemma 2,*

$$(4.8) \qquad \mathbf{E}\left(\operatorname{Per} A\right)^2 = (n!)^2 \left(\frac{N}{n^2}\right)^{2n} \exp\left\{-\left(\frac{n^2}{N}-1\right)\right\}$$

$$\times \left\{1 + O\left(n^{-1/4-\varepsilon}\right) + O\left(n^{-2\varepsilon}\right)\right\}.$$

Proof   According to (4.6) and (4.7) the representation

$$(4.9) \qquad \frac{(N)_{2n-r}}{(n^2)_{2n-r}} = \left(\frac{N}{n^2}\right)^{2n-r}$$

$$\times \exp\left\{-2\left(1-\frac{r}{2n}\right)^2\left(\frac{n^2}{N}-1\right)\right\}\left(1+O\left(\frac{1}{n^{2\varepsilon}}\right)\right)$$

holds for $r = 0, 1, \ldots, n$. We introduce the notations

$$(4.10) \qquad S_1 := \sum_{r=0}^{[n^{5/8}]} \frac{(N)_{2n-r}}{(n^2)_{2n-r}} \cdot \frac{1}{r!} \sum_{k=0}^{n-r} \frac{(-1)^k}{k!},$$

$$(4.11) \qquad S_2 := \sum_{r=[n^{5/8}]+1}^{n} \frac{(N)_{2n-r}}{(n^2)_{2n-r}} \cdot \frac{1}{r!} \sum_{k=0}^{n-r} \frac{(-1)^k}{k!}.$$

The quantity S_1 has the asymptotic expansion

$$(4.12) \quad S_1 = \sum_{r=0}^{[n^{5/8}]} \frac{e^{-1}}{r!} \left(\frac{N}{n^2}\right)^{2n-r} \exp\left\{-2\left(1-\frac{r}{N}\right)\left(\frac{n^2}{N}-1\right)\right\}$$

$$\times \left\{1 + O\left(n^{-1/4-\varepsilon}\right) + O\left(n^{-2\varepsilon}\right)\right\}.$$

By considering a Poisson distribution with parameter

$$\lambda = \frac{n^2}{N} \exp\left\{2\left(\frac{n}{N}-\frac{1}{n}\right)\right\}$$

we deduce that

$$(4.13) \qquad \sum_{r=0}^{[n^{5/8}]} \frac{\lambda^r e^{-\lambda}}{r!} \exp\left\{\lambda - \frac{n^2}{N}\right\} = 1 + O\left(n^{-2\varepsilon}\right),$$

which allows us to get the final estimate for S_1:

$$(4.14) \quad S_1 = \left(\frac{N}{n^2}\right)^{2n} \exp\left\{-\left(\frac{n^2}{N}-1\right)\right\}\left\{1 + O\left(n^{-1/4-\varepsilon}\right) + O\left(n^{-2\varepsilon}\right)\right\}.$$

The required estimate for S_2 is as follows:

$$(4.15) \qquad S_2 \le \sum_{r=[n^{5/8}]+1}^{n} \frac{(N)_{2n-r}}{(n^2)_{2n-r}} \frac{1}{r!} = \left(\frac{N}{n^2}\right)^{2n} O\left(n^{-n^{5/8}/8}\right).$$

Combining (4.14)–(4.15) with formula (4.3) we establish the conclusion of Lemma 3. $\square$

We now prove a theorem due to O'Neil [84].

Theorem 4.1 *Let $N = N(n)$ be an integer-valued function defined for $n = 1, 2, \ldots$ and satisfying the condition $N > n^{2/3+\varepsilon}$ for some $\varepsilon > 0$. If A is a matrix chosen at random from the set $\mathfrak{U}(n, N)$ then, for each $\gamma > 0$,*

$$(4.16) \qquad \lim_{n \to \infty} \mathbf{P}\left\{ \left| \frac{\mathrm{Per}\,A}{\mathbf{E}\,(\mathrm{Per}\,A)} - 1 \right| > \gamma \right\} = 1.$$

Proof Applying Chebyshev's inequality to $\mathrm{Per}\,A$ we have

$$\mathbf{P}(\,|\,\mathrm{Per}\,A - \mathbf{E}\,(\mathrm{Per}\,A)\,| \leq \delta\,\sqrt{\mathrm{Var}\,(\mathrm{Per}\,A)}\,) \geq 1 - \delta^{-2}, \qquad \delta > 0.$$

We take $\delta = n^\varepsilon$ and observe that, according to Lemmas 2 and 3, the asymptotic expansion

$$\sqrt{\mathrm{Var}\,(\mathrm{Per}\,A)} = \mathbf{E}\,(\mathrm{Per}\,A)\left[O\!\left(n^{-1/4-\varepsilon}\right) + O\!\left(n^{-2\varepsilon}\right)\right]$$

holds. As a result we obtain

$$(4.17) \quad \mathbf{P}\left\{ \mathrm{Per}\,A = \mathbf{E}\,(\mathrm{Per}\,A)\left(1 + O\!\left(n^{-1/4}\right) + O\!\left(n^{-\varepsilon}\right)\right) \right\} \geq 1 - n^{-2\varepsilon},$$

which completes the proof of Theorem 4.1.   □

The following modification of the result obtained was suggested by O'Neil [83]. Let $M^n(\mathbf{R}, \mathbf{S})$ be the class of square $(0, 1)$-matrices of order n whose row sums and column sums are given by vectors $\mathbf{R} = (r_1, r_2, \ldots, r_n)$ and $\mathbf{S} = (s_1, s_2, \ldots, s_n)$ respectively. We use the symbol $M^n(k, k)$ to indicate the class of square $(0, 1)$-matrices of order n, of which all row sums and column sums are equal to k.

A class $M^n(\mathbf{R}, \mathbf{S})$ is said to be a class of sparse matrices with associated function $f(n)$ if $r_i < f(n)$ and $s_i < f(n)$, $i = 1, 2, \ldots, n$.

A class $M^n(\mathbf{R}, \mathbf{S})$ is said to be a class of dual-sparse matrices with associated function $f(n)$ if $r_i > n - f(n)$ and $s_i > n - f(n)$, $i = 1, 2, \ldots, n$.

Let $A \in M^n(\mathbf{R}, \mathbf{S})$, where $M^n(\mathbf{R}, \mathbf{S})$ is the class of dual-sparse matrices with $f(n) = (\log n)^{1-\varepsilon}$, $\varepsilon > 0$, and let $L = L(A) = n^2 - \sum_{i=1}^n r_i$ be the number of zeros in A. Then

$$\mathrm{Per}\,A = n!\,e^{-L/n}\left(1 + O\!\left(n^{-1+\delta}\right)\right), \qquad\qquad n \to \infty,$$

for arbitrarily small $\delta > 0$ and sufficiently large n.

Further, let $M^n(\mathbf{R}, \mathbf{S})$ be a collection of sparse matrices of order n with $f(n) = (\log n)^{1/4-\varepsilon}$, $\varepsilon > 0$. Then, as $n \to \infty$, the number of matrices in this class can be computed by the formula

$$\left| M^n(R, S) \right| = \frac{\left(\sum r_i\right)!}{\prod r_i!\,\prod s_i!}$$

$$\times \exp\left\{ -\frac{\sum r_i(r_i - 1)\sum s_i(s_i - 1)}{2\left(\sum r_i\right)^2} \right\}\left(1 + O\!\left(n^{-1+\delta}\right)\right),$$

where all sums and products range from 1 to n and $\delta > 0$ is arbitrary.

The following assertion is valid.

Theorem 4.1' *Let A be a random equiprobable matrix of class $M^n(k,k)$. If $n \to \infty$ then*

(a) *for $2 \le k < (\log n)^{-\varepsilon+1/4}$ the mean value of Per A has the asymptotic representation*

$$\mathbf{E}\,(\mathrm{Per}) = \sqrt{\frac{2\pi n\,(k-1)}{e\,k}}\,\left(\frac{(k-1)^{k-1}}{k^{k-2}}\right)^n\left(1+o\!\left(n^{-1+\delta}\right)\right);$$

(b) *for $3 \le k < \log^{1/4} n$ the second moment of Per A has the asymptotic representation*

$$\mathbf{E}\,(\mathrm{Per})^2 = \frac{2\pi n\,(k-1)}{e\,k}\,\left(\frac{(k-1)^{k-1}}{k^{k-2}}\right)^{2n}$$

$$\times\left(1+\tfrac{1}{4}\,(k-1)^{-4}+O(k^{-6})\right)\left(1+O\!\left(n^{-1/2+\delta}\right)\right),$$

where $\delta > 0$ is arbitrary.

Using this theorem, O'Neil obtained the following statement.

Theorem 4.1'' *Let $k = g(n)$ be a monotonically increasing integer-valued function of n growing to infinity with n such that $g(n) < (\log n)^{-\varepsilon+1/4}$. Then for a random matrix $A \in M^n(k,k)$ the probability of the event that the ratio Per $A/\mathbf{E}\,(\mathrm{Per}\,A)$ is arbitrarily close to 1 tends to 1 as $n \to \infty$.*

2.4.2   Probability that the permanent of a random (0,1)-matrix with a prescribed number of units is positive

We will assume that

$$N = N_1 = N_1(n) = n\,\log n + C\,n + o(n)\,, \qquad n \to \infty\,,$$

where C is an arbitrary constant. Given this condition, it is not difficult to establish that the mean and variance of the permanent of a random equiprobable matrix $A \in \mathfrak{U}(n, N_1)$ satisfy, as $n \to \infty$, the asymptotic relations

(4.18) $\mathbf{E}\,(\mathrm{Per}\,A) = \exp\left\{n\left[\log\log n - 1 + \left(C - \tfrac{1}{2}\right)\dfrac{1}{\log n} + o\!\left(\dfrac{1}{\log n}\right)\right]\right\}$

and

(4.19) $\mathrm{Var}\,(\mathrm{Per}\,A) = \exp\left\{2n\,\log\log n + O(n)\right\}.$

These representations show that the methods used above are not applicable for studying the properties of the permanent of a random matrix $A \in \mathfrak{U}(n, N_1)$. For this reason, here we evaluate the probability $\mathbf{P}(n, N_1)$ that the permanent of a random matrix $A \in \mathfrak{U}(n, N_1)$ will be positive. The limiting behavior of the probability is described, as $n \to \infty$, by the following theorem due to Érdös and Rényi [36].

Theorem 4.2 *If* $N_1 = n \log n + C n + o(n)$, *where* C *is a constant then*

$$(4.20) \qquad \lim_{n \to \infty} \mathbf{P}(n, N_1) = \exp \left\{ -2 e^{-C} \right\}.$$

If $N_2 = n \log n + w(n) n + o(n)$ *then*

$$(4.21) \qquad \lim_{n \to \infty} \mathbf{P}(n, N_2) = 1 \qquad as \qquad w(n) \to +\infty,$$

$$(4.22) \qquad \lim_{n \to \infty} \mathbf{P}(n, N_2) = 0 \qquad as \qquad w(n) \to -\infty.$$

Proof First we prove equality (4.20). By the König–Frobenius theorem the condition Per $A = 0$ means that there exist k rows and $n - k - 1$ columns which cover all the units of A. We denote by $Q_k(n, N_1)$ the probability that a random matrix $A \in \mathfrak{U}(n, N_1)$ has k rows and $n - k - 1$ columns or $n - k - 1$ rows and k columns satisfying the above condition, and k is taken to be the minimal number possessing the property. It is clear that $0 \le k \le [(n - 1)/2]$ and

$$1 - \mathbf{P}(n, N_1) = \sum_{k=0}^{[(n-1)/2]} Q_k(n, N_1).$$

To prove equality (4.20) it suffices to establish that

$$(4.23) \qquad \lim_{n \to \infty} Q_0(n, N_1) = 1 - \exp \left\{ -2 e^{-C} \right\}$$

and

$$(4.24) \qquad \lim_{n \to \infty} \sum_{k=1}^{[(n-1)/2]} Q_k(n, N_1) = 0.$$

We observe that $1 - Q_0(n, N_1)$ is the probability that a random matrix A has no zero lines. Applying Theorem 4.1 with $\alpha = 1$ and $\gamma = C$ we see that the probability is equal to the probability that a random variable having Poisson distribution with parameter $\lambda = 2 \exp \{-C\}$ takes the value 0. Hence formula (4.23) follows.

Now we proceed to prove equality (4.24). Without loss of generality, one can represent any matrix A satisfying the condition Per $A = 0$ as follows:

$$(*) \qquad A = \begin{array}{c} \\ n - k - 1 \\ k + 1 \end{array} \overset{\begin{array}{cc} k & n - k \end{array}}{\left(\begin{array}{cc} U & V \\ W & O \end{array} \right)}.$$

Here all the units of A belong to the k rows and $n - k - 1$ columns, where k is taken to be the minimal number possessing this property. If the rows and columns are the first k rows and the first $n - k - 1$ columns of A then all the elements of the matrix O are equal to 0. We claim that each column of the submatrix W has at least two units. Indeed, let us assume that there exists a column of W, all but one of the elements of which are equal to 0. If we delete such a column from the coverage of 1s of the matrix A and add the row covering the positive element of the deleted column we obtain a coverage of units of the matrix A by $k - 1$ columns and $n - k$ rows. But this contradicts the minimality of k. Let us find an upper estimate for the probability $Q_k(n, N_1)$, $1 \leq k \leq [(n - 1)/2]$. With this aim, we consider an arbitrary matrix A satisfying $(*)$ with Per A $= 0$ and the zero matrix O of size $(k + 1) \times (n - k)$. First we note that, for any fixed k, there are at most

$$\binom{k + 1}{2}^k \binom{n(n - k - 1) + k(k - 1)}{N_1 - 2k}$$

such matrices. Indeed, the pairs of units in the columns of the submatrix W can be chosen in $\binom{k+1}{2}^k$ ways and for each such choice there are $n(n - k - 1) + k(k + 1) - 2k$ places for the remaining $N - 2k$ units. If we permute the columns and then the rows of any matrix A, leaving unchanged the internal order of the columns of U, the rows of W and of their complementary sets, we obtain from this A at most $\binom{n}{k}\binom{n}{n-k-1}$ matrices, each of which has zero permanent and the same value of k. Calculating the number of matrices transposed to the one just considered, we have

$$(4.25) \qquad Q_k(n, N_1) \leq 2 \binom{n}{k}\binom{n}{k + 1}\binom{k + 1}{2}^k$$

$$\times \binom{n(n - k - 1) + k(k - 1)}{N_1 - 2k}\binom{n^2}{N_1}^{-1}.$$

Hence, by Stirling's formula, we obtain, for $1 \leq k \leq [(n - 1)/2]$ and sufficiently large n,

$$(4.26) \qquad Q_k(n, N_1) \leq C_1^k \, n^{2k+1} \left(\frac{N_1}{n^2}\right)^{2k} \exp\left\{-\frac{k N_1}{2n} - \log n\right\}$$

$$\leq \left(\frac{C_2 \log^2 n}{\sqrt{n}}\right)^k,$$

where C_1 and C_2 are absolute positive constants. Using (4.26) we see that

$$(4.27) \qquad \sum_{k=1}^{[(n-1)/2]} Q_k(n, N_1) < \frac{C_2 \log^2 n}{\sqrt{n} - C_2 \log^2 n},$$

which completes the proof of equality (4.24).

By similar calculations one can show that

$$(4.28) \qquad \mathbf{P}(n, N_2) = \exp\left\{-2 e^{-w(n)}\right\}\left(1 + o(1)\right).$$

Hence, equalities (4.21) and (4.22) follow. □

2.4.3 Probability that the permanent of a random (0,1)-matrix with independent elements is positive

Let $A(p) = \| a_{ij} \|$, $i, j = 1, 2, \ldots, n$, be a random $(0, 1)$-matrix whose elements a_{ij} are independent random variables such that

$$\mathbf{P}\{a_{ij} = 1\} = p, \qquad \mathbf{P}\{a_{ij} = 0\} = q, \qquad p + q = 1.$$

We denote by $P_n(p)$ the probability that the permanent of $A(p)$ is positive.

Theorem 4.3 *If, for some constant C,*

$$p = \frac{\log n + C}{n} + o\left(\frac{1}{n}\right) \qquad as \qquad n \to \infty$$

then

$$\lim_{n \to \infty} P_n(p) = \exp\left\{-2\, e^{-C}\right\}.$$

Proof We prove the theorem by arguments similar to those used in the proof of Theorem 4.2. One has

$$1 - P_n(p) = \sum_{k=0}^{[(n-1)/2]} Q_{kn}(p),$$

where $Q_{kn}(p)$ is the probability that k rows and $k - n - 1$ columns or k columns and $n - k - 1$ rows can be taken which contain all the units of $A(p)$ and k is the minimal number with this property. By the inclusion–exclusion method we obtain

$$1 - Q_{0n}(p) = \sum_{l=0}^{2n} (-1)^l\, S_l,$$

where

$$S_0 := 1, \qquad S_l := \sum_{r=0}^{l} \binom{n}{r} \binom{n}{l-r} (1-p)^{n\, l - r\, (l-r)}, \qquad l = 1, 2, \ldots, 2n.$$

For each l we have the equality

$$\lim_{n \to \infty} S_l = \frac{2^l \exp\{-l\, C\}}{l!}.$$

Hence it follows that

$$\lim_{n \to \infty} \left(1 - Q_{0n}(p)\right) = \exp\left\{-2\, e^{-C}\right\}.$$

In addition, for $k = 1, 2, \ldots, [(n-1)/2]$ we have the estimates

$$Q_{kn}(p) \leq 2 \left(\frac{k+1}{2}\right)^k \binom{n}{k} \binom{n}{k+1} p^{2k} (1-p)^{(k+1)(n-k)},$$

which can be proved by the method of Theorem 4.2. Hence we conclude that

$$Q_{kn}(p) \leq \left(\theta \, \log^2 n / \sqrt{n} \right)^k, \qquad k = 1, 2, \ldots, [(n-1)/2],$$

where the constant θ depends solely on C. Therefore,

$$\lim_{n \to \infty} \sum_{k=1}^{[(n-1)/2]} Q_{kn}(p) = 0$$

and Theorem 4.2 follows. □

2.4.4 Disjoint permutations in random (0,1)-matrices

Denote by $\nu(A)$ the maximal number of pairwise disjoint permutations corresponding to the tuples of n noncollinear 1s of a square $(0,1)$-matrix of order n. Clearly, Per $A \geq \nu(A)$ and the condition $\nu(A) \geq 1$ is equivalent to the condition Per $A > 0$.

We fix a number N and consider the set of all n by n $(0,1)$-matrices with exactly N elements equal to 1. We specify the uniform distribution on the set of matrices by assigning the probability $\binom{n^2}{N}^{-1}$ to each such matrix. Put

$$P(n, N, r) := \mathbf{P}\{ \nu(A) \geq r \}, \qquad r = 1, 2, \ldots.$$

Clearly, $P(n, N, 1)$ is the probability of the event $\{\text{Per } A > 0\}$:

$$P(n, N, 1) = P(n, N) = \mathbf{P}\{ \text{Per } A > 0 \}.$$

We now formulate Theorems 4.4–4.6 due to Érdös and Rényi [36].

Theorem 4.4 *Let r be a fixed positive integer and*

$$N_r^* = n \, \log n + (r-1) n \, \log \log n + C n + o(n), \qquad n \to \infty,$$

for some constant C. Then

$$\lim_{n \to \infty} P(n, N_r^*, r) = \exp\left\{ -2 \, e^{-C/(r-1)!} \right\}.$$

Obviously, Theorem 4.2 is a particular case of Theorem 4.4 for $r = 1$.

Theorem 4.5 *Let r be a fixed positive integer and*

$$N_r = n \, \log n + (r-1) n \, \log \log n + n \, w(n) + o(n), \qquad n \to \infty,$$

where $w(n)$ tends arbitrarily slowly to infinity as $n \to \infty$. Then

$$\lim_{n \to \infty} P(n, N_r, r) = 1.$$

The following theorem is valid for random $(0,1)$-matrices whose elements are independent random variables taking the values 0 and 1 with probabilities p and q, respectively, $p + q = 1$.

Theorem 4.6 *Let r be a fixed positive integer and*

$$p = \frac{\log n + (r-1) \log \log n + w(n)}{n} + o\left(\frac{1}{n}\right), \qquad n \to \infty,$$

where $w(n)$ tends arbitrarily slowly to infinity as $n \to \infty$. Then

$$\lim_{n \to \infty} \mathbf{P}\{\nu(A) \geq r\} = 1.$$

In proving Theorem 4.5 we restrict ourselves to the consideration of the case where $r \geq 2$ since, for $r = 1$, the theorem has already been proved. Let us assume that $\nu(A) \leq r - 1$. Clearly, we can delete from each row and column of such a matrix A with N_r units $r - 1$ suitably selected units so that the permanent of the constructed matrix A$'$ will be equal to 0. Denote by $Q_1(n, r)$ the probability that the resulting matrix A$'$ contains a zero line, and by $Q_2(n, r)$ the probability that Per A$' = 0$ and A$'$ has no zero lines.

It can easily be seen that if a row (column) of A$'$ consists of zeros only, then the corresponding row (column) of A contains at most $r - 1$ units. Conversely, if A contains such a row or column then $\nu(A) \leq r - 1$. Thus, $Q_1(n, r)$ is the probability of the event that A contains at least one row or column which has at most $r - 1$ units. Hence we find that

$$Q_1(n, r) \leq 2n \sum_{j=0}^{r-1} \binom{n}{j} \binom{n^2 - n}{N_r - j} \Big/ \binom{n^2}{N_r}.$$

Estimating the right-hand side of the preceding inequality from above we see that

$$Q_1(n, r) \leq 2rn \left(\frac{N_r}{n}\right)^{r-1} \exp\left\{-\frac{N_r}{n}\right\} (1 + o(1)).$$

Recalling the definition of N_r we finally obtain

$$Q_1(n, r) = O\left(e^{-w(n)}\right).$$

Now we proceed to consider the second case: Per A$' = 0$ and A$'$ has no zero lines. Let k be the least number such that one can find in A$'$ either k columns and $n - k - 1$ rows, or k rows and $n - k - 1$ columns, which contain all the units of A$'$ and $1 \leq k \leq [(n-1)/2]$. For the sake of definiteness we assume that all the units of A$'$ can be covered by k columns and $n - k - 1$ rows. By symmetry the coverage probability of units of A$'$ by k rows and $n - k - 1$ columns is the same. Arguments similar to those used in the proof of Theorem 4.2 show that there exists a submatrix W$'$ of the matrix A$'$

with $k+1$ rows and k columns such that each column of W' contains at least two units. Let W be the corresponding submatrix of A'. Denote by δ_k the probability that A contains a $k+1$ by k submatrix W such that each column of W contains at least two units, and the submatrix O of A, formed by the same rows as W and by those columns which do not intersect W, contains at most $r-1$ units in each row. It follows that

$$Q_2(n,r) \le 2 \sum_{k=1}^{[(n-1)/2]} \delta_k.$$

It is clear that

$$\delta_k \le \binom{n}{k}\binom{n}{k+1}\binom{k+1}{2}^k$$

$$\times \sum_{j=0}^{(k+1)(r-1)} \binom{(k+1)(n-k)}{j}\binom{n(n-k-1)+k(k-1)}{N_r-2k-j} \Big/ \binom{n^2}{N_r}.$$

By arguments similar to those used in the proof of Theorem 4.2, one can state that

$$Q_2(n,r) = o(1).$$

Now the theorem immediately follows from the obvious relation

$$1 - P(n, N_r, r) = Q_1(n,r) + Q_2(n,r). \qquad \square$$

2.5   The mean value of the permanent of a double stochastic random matrix

2.5.1   The exact formula

We consider the class of double stochastic matrices of the form

$$A = s^{-1}\left(\Pi_1 + \Pi_2 + \cdots + \Pi_s\right),$$

where $\Pi_1, \Pi_2, \ldots, \Pi_s$ are permutation matrices of order n. We assume that the $(n!)^s$ distinct tuples of permutation matrices are ordered in a certain way and specify the uniform distribution on the elements of the set. Our aim is to show that, under these conditions, the mean value of the permanent of a random matrix A can be computed by the formula [123]:

(5.1) $\mathbf{E}\,(\mathrm{Per}\,A) = s^{-n}\,(n!)^{2-s}$

$$\times \sum_{r_1+r_2+\cdots+r_s=n} \frac{(n-r_1)!\,(n-r_2)\cdots(n-r_s)!}{r_1!\,r_2!\cdots r_s!},$$

where summation is taken over all nonnegative integer-valued solutions of the equation $r_1 + r_2 + \cdots + r_s = n$. We have

$$\mathbf{E}\left(\mathrm{Per}\,A\right) = (n!)^{-s} \sum_{\Pi_1, \ldots, \Pi_s} \mathrm{Per}\left(s^{-1}\left(\Pi_1 + \Pi_2 + \cdots + \Pi_s\right)\right),$$

where summation is accomplished over all $(n!)^s$ tuples of permutation matrices $\Pi_1, \Pi_2, \ldots, \Pi_s$. Letting $\delta_{ii} = 1$ and $\delta_{ij} = 0$, $i \neq j$, we represent a permutation matrix Π_k, $k = 1, 2, \ldots, s$, in the form

$$\Pi_k = \left\| \delta_{\varphi_k(i),j} \right\|, \qquad\qquad i, j = 1, 2, \ldots, n,$$

where φ_k is the permutation corresponding to Π_k. Obvious calculations show that

$$\sum_{\Pi_1, \ldots, \Pi_s} \mathrm{Per}\left(s^{-1}\left(\Pi_1 + \Pi_2 + \cdots + \Pi_s\right)\right) = s^{-n} \sum_{\varphi_1, \ldots, \varphi_s}$$

$$\times \sum_{(j_1, \ldots, j_n)} \left(\delta_{\varphi_1(1),j_1} + \cdots + \delta_{\varphi_s(1),j_1}\right) \cdots \left(\delta_{\varphi_1(n),j_n} + \cdots + \delta_{\varphi_s(n),j_n}\right),$$

where the external summation is extended over all admissible tuples of s permutations of order n and the internal one is taken over all permutations of order n. For a fixed permutation $(j_1, j_2, \ldots, j_n)$ we calculate the product of the corresponding expressions in brackets. After the multiplication we consider an arbitrary term of the obtained sum. The term is a product of n factors, each of which is a summand of the corresponding sum in brackets. Assume that the product is comprised by r_1 first summands, r_2 second summands, $\ldots$, and r_s sth summands of the expressions in brackets, where $r_1 + r_2 + \cdots + r_s = n$, $r_i \geq 0$, $i = 1, 2, \ldots, s$. In order that the corresponding product be positive it is necessary and sufficient that, for all $k = 1, 2, \ldots, s$, exactly r_k images of the permutation φ_k coincide with certain images of the permutation γ: $\gamma(i) = j_i$, $i = 1, 2, \ldots, n$, and the images be different for distinct permutations. Choose a permutation γ and consider the set of all tuples of permutations $\varphi_1, \varphi_2, \ldots, \varphi_s$ in which r_1 images of φ_1, r_2 images of φ_2, $\ldots$, r_s images of φ_s are fixed. The number of such tuples is

$$(n - r_1)! \, (n - r_2)! \cdots (n - r_s)!.$$

Since the images in distinct permutations are different by our hypothesis, we see that, for any fixed γ, the number of ways of choosing the images for the permutations $\varphi_1, \varphi_2, \ldots, \varphi_s$ is given by the formula

$$\binom{n}{r_1}\binom{n-r_1}{r_2} \cdots \binom{n - r_1 - \cdots - r_{s-1}}{r_s} = \frac{n!}{r_1! \, r_2! \cdots r_s!}.$$

The resulting expressions are independent of γ and we finally obtain

$$\sum_{\Pi_1, \dots, \Pi_s} \mathrm{Per} \left(s^{-1} \left(\Pi_1 + \cdots + \Pi_s \right) \right)$$

$$= (n!)^2 \ s^{-n} \sum_{r_1 + \cdots + r_s = n} \frac{(n - r_1)! \ \cdots \ (n - r_s)!}{r_1! \ \cdots \ r_s!} .$$

Obvious transforms of this relation lead to (5.1).

Formula (5.1) easily implies the estimate

$$\mathbf{E} \left(\mathrm{Per} \, A \right) \geq n! \ n^{-n}.$$

2.5.2   Asymptotic behavior for large s

We now show that, for a fixed n, the equality

(5.2) $$\lim_{s \to \infty} \mathbf{E} \left(\mathrm{Per} \, A \right) = n! \ n^{-n}$$

holds (see [123]). Observe that

$$\sum_{r_1 + \cdots + r_s = n} \frac{(n - r_1)! \ \cdots \ (n - r_s)!}{r_1! \ \cdots \ r_s!} = \mathrm{coef}_{z^n} \left(\sum_{k=0}^{n} \frac{(n - k)!}{k!} \ z^k \right)^s .$$

We derive by appeal to Cauchy's formula

(5.3) $$\mathbf{E} \left(\mathrm{Per} \, A \right) = s^{-n} \ (n!)^{2-s} \ \frac{1}{2\pi i} \oint_C \left(\sum_{k=0}^{n} \frac{(n - k)!}{k!} \ z^k \right)^s \frac{dz}{z^{n+1}} ,$$

where C is an arbitrary closed contour enclosing the origin in the complex plane z. We make a change of variables in the integral by setting $w = z \, s$. Let C' be the image of the contour C under such a transform. We then obtain

(5.4) $$\mathbf{E} \left(\mathrm{Per} \, A \right) = \frac{(n!)^2}{2\pi i} \oint_{C'} \left\{ \sum_{k=0}^{n} \frac{(n - k)!}{k! \, n!} \left(\frac{w}{s} \right)^k \right\}^s \frac{dw}{w^{n+1}} .$$

Let us show that

(5.5) $$\lim_{s \to \infty} \left\{ \sum_{k=0}^{n} \frac{(n - k)!}{k! \, n!} \left(\frac{w}{s} \right)^k \right\}^s = e^{w/n}$$

uniformly for $w \in C'$. We have

$$s \, \log \left(1 + \frac{w}{n \, s} + R_s(w) \right) = \frac{w}{n} + \frac{\bar{R}_s(w)}{s} ,$$

where the remainder terms $R_s(w)$ and $\bar{R}_s(w)$ are uniformly bounded on the contour C'. Hence, equality (5.5) follows. Letting $s \to \infty$ on both sides of (5.4) we conclude that

$$(5.6) \qquad \lim_{s \to \infty} \mathbf{E}\,(\operatorname{Per} A) = \frac{(n!)^2}{2\pi i} \oint_{C'} e^{w/n} \frac{dw}{w^{n+1}}.$$

The integrand on the right-hand side of (5.6) has a unique singular point $w_0 = 0$ within C'. Applying Cauchy's residue theorem yields

$$(5.7) \qquad \oint_{C'} e^{w/n} \frac{dw}{w^{n+1}} = 2\pi i \operatorname*{res}_{w=0} \left\{ \frac{e^{w/n}}{w^{n+1}} \right\} = \frac{2\pi i}{n^n n!},$$

where the symbol $\operatorname{res} f(z)$ stands for residue of $f(z)$. Now formula (5.2) follows from (5.6) and (5.7).

2.5.3 Asymptotic behavior for large n

Here we study the asymptotic behavior of $\mathbf{E}\,(\operatorname{Per} A)$ when s is fixed and $n \to \infty$. The following identity is valid:

$$\left[x_1 \cdots x_s \left(\frac{1}{x_1} + \cdots + \frac{1}{x_s} \right) \right]^n$$

$$= \sum_{r_1 + \cdots + r_s = n} \frac{n!}{r_1! \cdots r_s!}\, x_1^{n-r_1} \cdots x_s^{n-r_s}.$$

Multiplying both sides of the equality by $\exp\{-(x_1 + \cdots + x_s)\}$ and integrating the resulting relation with respect to $x_1, x_2, \ldots, x_s$ from 0 to ∞ we see that

$$\int_0^\infty \cdots \int_0^\infty \left[x_1 \cdots x_s \left(\frac{1}{x_1} + \cdots + \frac{1}{x_s} \right) \right]^n \exp\{-(x_1 + \cdots + x_s)\}\, dx_1 \ldots dx_s$$

$$= n! \sum_{r_1 + \cdots + r_s = n} \frac{(n - r_1)! \cdots (n - r_s)!}{r_1! \cdots r_s!}.$$

Recalling formula (5.1) we obtain

$$\mathbf{E}\,(\operatorname{Per} A) = \frac{1}{s^n (n!)^{s-1}} \int_0^\infty \cdots \int_0^\infty \left[x_1 \cdots x_s \left(\frac{1}{x_1} + \cdots + \frac{1}{x_s} \right) \right]^n$$

$$\times \exp\{-(x_1 + \cdots + x_s)\}\, dx_1 \ldots dx_s.$$

One can check by means of standard methods that the integrand attains its maximum at the point

$$x_1^0 = \cdots = x_s^0 = n(1 - s^{-1})$$

only. We set $\alpha = 1 - s^{-1}$ and substitute the variables by the formulae $x_i = n(u_i + \alpha)$, $i = 1, 2, \dots, s$. Applying Stirling's formula to estimate $n!$ we have

$$\mathbf{E}\,(\mathrm{Per}\,A) = \frac{n^{(s+1)/2}}{s^n\,(2\pi)^{(s-1)/2}} \int\limits_{-\alpha}^{\infty} \cdots \int\limits_{-\alpha}^{\infty} \Big[(u_1 + \alpha) \cdots (u_s + \alpha)$$

$$\times \left(\frac{1}{u_1 + \alpha} + \cdots + \frac{1}{u_s + \alpha}\right)$$

$$\times \exp\{-(u_1 + \cdots + u_s)\}\Big]^n du_1 \dots du_s \left(1 + o(1)\right).$$

The integrand has a unique maximum at the point $u_1^0 = \cdots = u_s^0 = 0$. Put

$$f(u_1, u_2, \dots, u_s) := \sum_{i=1}^{s} \log\,(u_i + \alpha) + \log \sum_{i=1}^{s} (u_i + \alpha)^{-1} - \sum_{i=1}^{s} u_i\,.$$

Using this notation one can write

$$\mathbf{E}\,(\mathrm{Per}\,A) = \frac{n^{(s+1)/2}}{(2\pi)^{(s-1)/2}\,s^n}$$

$$\times \int\limits_{-\alpha}^{\infty} \cdots \int\limits_{-\alpha}^{\infty} \exp\{nf(u_1, u_2, \dots, u_s)\}\,du_1\,du_2\,\dots\,du_s\,.$$

We denote

$$\varphi(u_1, u_2, \dots, u_s) := f(u_1, u_2, \dots, u_s) - f(0, 0, \dots, 0)$$

and observe that

$$f(0, 0, \dots, 0) = \log\left(s\,(1 - s^{-1})^{s-1}\right).$$

With these equalities in view we obtain

$$\mathbf{E}\,(\mathrm{Per}\,A) = \frac{n^{(s+1)/2}}{(2\pi)^{(s-1)/2}} \left(1 - \frac{1}{s}\right)^{n\,(s-1)}$$

$$\times \int\limits_{-\alpha}^{\infty} \cdots \int\limits_{-\alpha}^{\infty} \exp\{n\,\varphi(u_1, u_2, \dots, u_s)\}\,du_1\,du_2\,\dots\,du_s\,.$$

Substituting the variables by the formulae $y_1 = \sqrt{n}\, u_1, \ldots, y_s = \sqrt{n}\, u_s$ we get the following relation for the mean value of the permanent of a random matrix:

$$\mathbf{E}\,(\mathrm{Per}\,A) = \frac{n^{1/2}}{(2\pi)^{(s-1)/2}} \left(1 - \frac{1}{s}\right)^{n\,(s-1)}$$

$$\times \int\limits_{-\alpha\sqrt{n}}^{\infty} \cdots \int\limits_{-\alpha\sqrt{n}}^{\infty} \exp\left\{n\,\varphi\left(\frac{y_1}{\sqrt{n}}, \ldots, \frac{y_s}{\sqrt{n}}\right)\right\} dy_1 \ldots dy_s\,.$$

We represent the exponent of the integrand as a Maclaurin series

$$n\,\varphi\left(\frac{y_1}{\sqrt{n}}, \ldots, \frac{y_s}{\sqrt{n}}\right) = -\frac{1}{2}\sum_{i=1}^{s}\sum_{j=1}^{s} c_{ij}\, y_i\, y_j + O\left(\frac{y_1^3 + \cdots + y_s^3}{\sqrt{n}}\right),$$

where

$$c_{ij} := -\frac{\partial}{\partial u_i}\,\frac{\partial}{\partial u_j}\,\varphi(u_1, \ldots, u_s)\,\Big|_{u_1 = \cdots = u_s = 0}.$$

The integrand decreases exponentially and so

$$\mathbf{E}\,(\mathrm{Per}\,A) = \frac{n^{1/2}}{(2\pi)^{(s-1)/2}} \left(1 - \frac{1}{s}\right)^{n\,(s-1)}$$

$$\times \int\limits_{-n^{1/9}}^{n^{1/9}} \cdots \int\limits_{-n^{1/9}}^{n^{1/9}} \exp\left\{-\tfrac{1}{2}\sum_{i=1}^{s}\sum_{j=1}^{s} c_{ij}\, y_i\, y_j\right\} dy_1 \ldots dy_s \left(1 + o(1)\right),$$

where $o(1)$ decreases exponentially as $n \to \infty$. We can rewrite the expression as

$$\mathbf{E}\,(\mathrm{Per}\,A) = \frac{n^{1/2}}{(2\pi)^{(s-1)/2}} \left(1 - \frac{1}{s}\right)^{n\,(s-1)} J \left(1 + o(1)\right),$$

where, as before, $o(1)$ decreases exponentially as $n \to \infty$ and

$$J := \int\limits_{-\infty}^{\infty} \cdots \int\limits_{-\infty}^{\infty} \exp\left\{-\tfrac{1}{2}\sum_{j=1}^{s} c_{ij}\, y_i\, y_j\right\} dy_1 \ldots dy_s\,.$$

The quadratic form

$$\theta(y_1, y_2, \ldots, y_s) := \tfrac{1}{2}\sum_{i=1}^{s}\sum_{j=1}^{s} c_{ij}\, y_i\, y_j$$

is positive definite and, therefore, there exists an orthogonal linear transform of its variables leading to a sum of squares. Exploiting this fact we present the integral J as

$$J = \int\limits_{-\infty}^{\infty} \cdots \int\limits_{-\infty}^{\infty} \exp\left\{-\tfrac{1}{2} \sum_{j=1}^{s} \lambda_j v_j^2 \right\} dv_1 \ldots dv_s,$$

where $\lambda_1, \lambda_2, \ldots, \lambda_s$ are the characteristic roots of the matrix $C = \| c_{ij} \|$, $i, j = 1, 2, \ldots, s$. We obtain

$$\int\limits_{-\infty}^{\infty} \exp\left\{-\tfrac{1}{2} \lambda_i v_i^2 \right\} dv_i = \left(\frac{2\pi}{\lambda_i}\right)^{1/2}.$$

Thus, denoting by $\det C$ the determinant of the matrix C, we see that

$$J = \left(\frac{(2\pi)^s}{\lambda_1 \lambda_2 \cdots \lambda_s}\right)^{1/2} = \left(\frac{(2\pi)^s}{\det C}\right)^{1/2}.$$

Therefore,

$$\mathbf{E}\left(\operatorname{Per} A\right) = \left(\frac{2\pi n}{\det C}\right)^{1/2}\left(1 - \frac{1}{s}\right)^{n\,(s-1)}\left(1 + o(1)\right).$$

It remains to calculate $\det C$. Direct computations show that

$$c_{ij} = \begin{cases} 1 & \text{if } i = j, \\ (s-1)^{-2} & \text{if } i \neq j. \end{cases}$$

Subtracting the first column of $\det C$ from the remaining columns and then adding all the remaining rows to the first row of the determinant, we obtain the determinant of a triangular matrix. Calculating the product of the main diagonal elements of the matrix we conclude that

$$\det C = \left(1 - \frac{2}{s}\right)^{s-1}\left(1 - \frac{1}{s}\right)^{1-2s},$$

which leads to the following final expression for $s > 2$ (see [123]):

(5.8) $$\mathbf{E}\left(\operatorname{Per} A\right) = (2\pi n)^{1/2}\left(1 - \frac{1}{s}\right)^{n(s-1)+s-1/2}$$

$$\times \left(1 - \frac{2}{s}\right)^{-(s-1)/2}\left(1 + o(1)\right).$$

For $s = 2$ we derive from (5.1) that

$$\mathbf{E}\left(\operatorname{Per} A\right) = \frac{1}{2^n} \sum_{r=0}^{n} \frac{(n-r)!\, r!}{r!\, (n-r)!} = \frac{n+1}{2^n}.$$

The case $s = 1$ is trivial.

CHAPTER 3

Probabilistic Problems in
the General Combinatorial Scheme

3.0 Introduction

The probabilistic problems considered in this chapter are formulated in the framework of the general combinatorial scheme. A detailed description of the scheme is given in Chapter 5 of the monograph [111]. Below we list a number of combinatorial results related to the general combinatorial scheme that will be needed in this chapter.

Let σ be a *configuration*, that is, a mapping of a set $\mathbf{X} = \{1, 2, \ldots, m\}$ into a set $\mathbf{A} = \{a_1, a_2, \ldots, a_n\}$. The configuration σ is said to have the *primary specification* $\left[a_1^{\alpha_1} a_2^{\alpha_2} \ldots a_n^{\alpha_n} \right]$, $\alpha_1 + \alpha_2 + \cdots + \alpha_n = m$, if a_j is the image of exactly α_j elements of $\mathbf{X}$ under the mapping σ, $j = 1, 2, \ldots, n$. The *secondary specification* of σ is the expression $\left[\left[0^{\beta_1} 1^{\beta_2} 2^{\beta_3} \ldots m^{\beta_m} \right]\right]$, $1\,\beta_1 + 2\,\beta_2 + \cdots + m\,\beta_m = m$, where β_0 is the number of zeros, β_1 is the number of units, and so on, among $\alpha_1, \alpha_2, \ldots, \alpha_n$.

We consider permutation groups G and H acting on the sets $\mathbf{X}$ and $\mathbf{A}$ and having orders m and n respectively. We define a *GH-equivalence* relation on the set of configurations $\sigma \in \mathbf{A}^{\mathbf{X}}$ according to which $\sigma \sim \sigma'$ if there exist permutations $g \in G$ and $h \in H$ such that $g \circ \sigma * h = \sigma'$, where $g \circ \sigma * h(x) = g\big(\sigma(h(x))\big)$ for any $x \in \mathbf{X}$, that is, one first acts at x by the permutation h, then finds the image of $h(x)$ under the configuration σ, and, finally, acts at $\sigma\big(h(x)\big)$ by the permutation g.

In this chapter we show that the description of combinatorial configurations is reduced to constructing a factor-set with respect to a *GH-*equivalence under the appropriate choice of the permutation groups G and H. Four cases of the choice of permutation groups are of special interest, where we choose for G and H either the trivial groups or the symmetric groups of the correspondings orders. For these cases, termed particular cases

of the general combinatorial scheme, we give a method for constructing the
generating functions that enumerate elements of the corresponding factor-
sets by the values of primary and secondary specifications. Depending on
a particular choice of G and H in the general combinatorial scheme, the
elements of the corresponding factor-sets are called m-samples in a commutative (noncommutative) symmetric (nonsymmetric) n-basis. Below, we
give the generating functions enumerating m-samples in the particular cases
mentioned and a number of formulae that will be required for solving various
probabilistic problems related to the general combinatorial scheme.

3.0.1 Commutative nonsymmetric n-basis

This case of the general combinatorial scheme corresponds to a choice of
the symmetric group $\mathbb{S}_m$ as group G and the trivial group as group H. For
$1 \le j \le n$, let Λ_j be a subsequence of the sequence $\mathbb{N}_0 = \{0, 1, 2, \dots\}$.
Denote by $\Lambda = (\Lambda_1, \Lambda_2, \dots, \Lambda_n)$ the ordered tuple of the chosen subse-
quences and by $C_{nm}(\Lambda)$ the number of m-samples of a commutative non-
symmetric n-basis with primary specification $\left[a_1^{\alpha_1} a_2^{\alpha_2} \dots a_n^{\alpha_n} \right]$ such that
$\alpha_j \in \Lambda_j$, $j = 1, 2, \dots, n$. We have

$$(0.1) \qquad \sum_{m=0}^{\infty} C_{nm}(\Lambda) \, t^m = \prod_{j=1}^{n} \sum_{\alpha_j \in \Lambda_j} t^{\alpha_j}.$$

One can deduce from (0.1) that

$$(0.2) \qquad \sum_{m=ks}^{\infty} C_{nm}^{(k)}(s) \, t^m = t^{ks} \left(\frac{1}{1-t} - t^s \right)^{n-k},$$

where $C_{nm}^{(k)}(s)$ is the number of m-samples of the commutative nonsymmetric n-basis in which k fixed elements (and k only) occur exactly s times.
Putting $s = 0$ we see that

$$(0.3) \qquad C_{nm}^{(k)}(0) = \binom{m-1}{n-k-1}.$$

Let $Q_{nm}^{(k)}(s)$ be the number of m-samples in the n-basis for each of which
s fixed elements $a_{i_1}, a_{i_2}, \dots, a_{i_s}$ occur $k_{i_1}, k_{i_2}, \dots, k_{i_s}$ times,
respectively, $k_{i_1} + k_{i_2} + \cdots + k_{i_s} = k$, and let $\bar{Q}_{nm}^{(k)}(s)$ be the number
of m-samples satisfying the conditions imposed above such that each of the
remaining elements occurs at least once. From equality (0.1) we find that

$$(0.4) \qquad \sum_{m=k}^{\infty} Q_{nm}^{(k)}(s) \, t^m = t^k \, (1-t)^{-(n-s)}.$$

Hence it follows that

$$(0.5) \qquad Q_{nm}^{(k)}(s) = \binom{n+m-s-k-1}{m-k}.$$

Similar arguments show that

$$(0.6) \qquad \bar{Q}_{nm}^{(k)}(s) = \binom{m-k-1}{n-s-1}.$$

Let us consider the set of sequences of length m, each of which has m_0 zeros and m_1 units, $m_0 + m_1 = m$. Denote by $N_m^{(0)}(k)$ the number of such sequences with k series of zeros, and by $M_m(2k)$ the number of such sequences with $2k$ series of both kinds provided that elements of the sequences are located on a circle. The following formulae are valid:

$$(0.7) \qquad N_m^{(0)}(k) = \binom{m_0-1}{k-1}\binom{m_1+1}{k},$$

$$(0.8) \qquad M_m(2k) = \frac{m}{k}\binom{m_0-1}{k-1}\binom{m_1-1}{k-1}.$$

3.0.2 Noncommutative nonsymmetric n-basis

This case of the general combinatorial scheme is specified by the condition that the groups G and H are trivial and so there are n^m equivalence classes called m-samples in a noncommutative nonsymmetric n-basis. Assuming that Λ has the same meaning as in subsection 3.0.1, we denote by $D_{nm}(\Lambda)$ the number of m-samples with primary specifications $\left[a_1^{\alpha_1} a_2^{\alpha_2} \ldots a_n^{\alpha_n}\right]$ such that $\alpha_j \in \Lambda_j$, $j = 1, 2, \ldots, n$. Then

$$(0.9) \qquad \sum_{m=0}^{\infty} D_{nm}(\Lambda)\frac{t^m}{m!} = \prod_{j=1}^{n} \sum_{\alpha_j \in \Lambda_j} \frac{t^{\alpha_j}}{\alpha_j!}.$$

By means of the generating function we obtain

$$(0.10) \qquad \sum_{m=ks}^{\infty} D_{nm}^{(k)}(s)\frac{t^m}{m!} = \left(\frac{t^s}{s!}\right)^k \left(e^t - \frac{t^s}{s!}\right)^{n-k},$$

where $D_{nm}^{(k)}(s)$ is the number of such samples of the noncommutative nonsymmetric n-basis in which each of k fixed elements of the n-basis (and only the k elements) occurs exactly s times. It follows from (0.10) that

$$(0.11) \qquad D_{nm}^{(k)}(0) = \Delta^{n-k} 0^m = \sum_{j=0}^{n-k} (-1)^j \binom{n-k}{j}(n-k-j)^m,$$

where $\Delta^n 0^m$ is the nth finite difference of x^m at the point $x = 0$, usually termed Morgan's number. Denoting by $Q(n, m, k)$ the number of m-samples in which a fixed element of the n-basis occurs k times, we deduce from (0.9) that

$$(0.12) \qquad \sum_{m=k}^{\infty} Q(n, m, k)\, \frac{t^m}{m!} = \frac{t^k}{k!}\, e^{(n-1)t}.$$

3.0.3 Noncommutative symmetric n-basis

If G is the trivial group and H is the symmetric group of order n, we call the elements of the corresponding factor-sets in the general combinatorial scheme m-samples of a noncommutative symmetric n-basis. With each such m-sample we associate a partition of an m-element set into at most n blocks.

Let $\Lambda = (\Lambda_1, \Lambda_2, \ldots)$, $\Lambda_j \subseteq \mathbb{N}_0 = \{0, 1, \ldots\}$, and let $T_{nm}(\Lambda)$ be the number of partitions of an m-element set into n blocks in each of which the number of blocks containing exactly j elements satisfies the condition $\beta_j \in \Lambda_j$, $j = 1, 2, \ldots$. Put

$$(0.13) \qquad T_m(\Lambda) := \sum_{n=0}^{m} T_{nm}(\Lambda), \qquad\qquad T_0(\Lambda) := T_{00}(\Lambda) = 1$$

and denote by $T_m(\Lambda; l, k)$ the number of such partitions containing k blocks of size l. The following relations are valid for the corresponding generating functions:

$$(0.14) \qquad \sum_{m=0}^{\infty} T_m(\Lambda)\, \frac{t^m}{m!} = \prod_{j=1}^{\infty} \sum_{\beta_j \in \Lambda_j} \left(\frac{t^j}{j!}\right)^{\beta_j} \frac{1}{\beta_j!},$$

$$(0.15) \qquad \sum_{m=0}^{\infty} \sum_{n=0}^{m} T_{nm}(\Lambda)\, \frac{t^m}{m!}\, x^n = \prod_{j=1}^{\infty} \sum_{\beta_j \in \Lambda_j} \left(\frac{xt^j}{j!}\right)^{\beta_j} \frac{1}{\beta_j!},$$

$$(0.16) \quad \sum_{m=0}^{\infty} \sum_{k=0}^{[m/l]} T_m(\Lambda; l, k)\, \frac{t^m}{m!}\, x^n = \sum_{\beta_l \in \Lambda_l} \left(\frac{x t^l}{l!}\right)^{\beta_l}$$

$$\times \frac{1}{\beta_l!} \prod_{\substack{j=1 \\ j \neq l}}^{\infty} \sum_{\beta_j \in \Lambda_j} \left(\frac{t^j}{j!}\right)^{\beta_j} \frac{1}{\beta_j!}.$$

In particular, if $\Lambda_j = \mathbb{N}_0 = \{0, 1, \ldots\}$, $j = 1, 2, \ldots$, then

$$T_m = T_m(\mathbb{N}_0), \quad T_{mn} = T_{mn}(\mathbb{N}_0), \quad T_m(l, k) = T_m(\mathbb{N}_0; l, k)$$

and

$$(0.17) \qquad \sum_{m=0}^{\infty} T_m \frac{t^m}{m!} = \exp\{e^t - 1\},$$

$$(0.18) \qquad \sum_{m=0}^{\infty} \sum_{n=0}^{m} T_{mn} \frac{t^m}{m!} x^n = \exp\{x(e^t - 1)\},$$

$$(0.19) \quad \sum_{m=0}^{\infty} \sum_{k=0}^{[m/l]} T_m(l, k) \frac{t^m}{m!} x^k = \exp\left\{(x-1)\frac{t^l}{l!} + e^t - 1\right\}.$$

The numbers T_m are termed Bell's numbers. We have

$$(0.20) \qquad T_{m+1} = \sum_{k=0}^{m} \binom{m}{k} T_{m-k}, \qquad\qquad T_0 = 1,$$

whereas $T_{mn} = \sigma(m,n)$ and $\sigma(m,n)$, $m \le n$, are Stirling's numbers of the second kind.

Let T_{mn}^A be the number of partitions of an m-element set into n blocks whose sizes are elements of the sequence $A \subseteq \mathbb{N} = \{1, 2, \ldots\}$, and let T_m^A and $T_m^A(k, l)$ be, respectively, the number of distinct partitions of the m-element set into blocks and the number of partitions of the set with k blocks of size $l \in A$. Setting $A(t) := \sum_{j\in A} t^j/j!$ we deduce from (0.14)–(0.16) that

$$(0.21) \qquad \sum_{m=0}^{\infty} T_m^A \frac{t^m}{m!} = e^{A(t)},$$

$$(0.22) \qquad \sum_{m=0}^{\infty} \sum_{n=0}^{m} T_{mn}^A \frac{t^m}{m!} x^n = e^{xA(t)},$$

$$(0.23) \quad \sum_{m=0}^{\infty} \sum_{k=0}^{[m/l]} T_m^A(k, l) \frac{t^m}{m!} x^k = \exp\left\{(x-1)\frac{t^l}{l!} + A(t)\right\}.$$

The mth Bell number satisfies Dobinski's formula

$$(0.24) \qquad T_m = \frac{1}{e} \sum_{k=0}^{\infty} \frac{k^m}{k!},$$

and, as $m \to \infty$, the asymptotic relation (see [79])

$$(0.25) \qquad T_m = \frac{1}{\sqrt{r+1}} \exp\left\{m\left(r + \frac{1}{r} - 1\right) - 1\right\}\left(1 + o(1)\right),$$

where r is the unique positive solution of the equation $r\,e^2 = m$ and, as $m \to \infty$,

$$(0.26) \quad r = \log m - \log\log m + \sum_{k=0}^{\infty} \sum_{l=0}^{\infty} C_{kl} (\log\log m)^{l+1} (\log m)^{-k-l-1},$$

where the constants C_{kl} are independent of m.

3.1 Probability distributions
for a commutative nonsymmetric n-basis

We specify the uniform distribution on the set of m-samples of a noncommutative nonsymmetric n-basis by prescribing to each m-sample the probability $\binom{n+m-1}{m}^{-1}$. We will study some random variables defined on a probability space and investigate their exact and limiting (as n and m tend to infinity) distributions.

3.1.1 Distribution of the number of elements of an n-basis occurring a given number of times

We denote by $\xi_{nm}(s)$ the number of elements of an n-basis occurring exactly s times in a random m-sample. Considering a random allocation of m identical objects into n unlike cells we see that $\xi_{nm}(s)$ is the number of cells containing s objects. It is clear that

$$(1.1) \qquad \mathbf{P}\{\xi_{nm}(s) = k\} = \binom{n}{k}\binom{n+m-1}{m}^{-1} C_{nm}^{(k)}(s),$$

$$k = 0, 1, \dots, n,$$

where $C_{nm}^{(k)}(s)$ is the number of m-samples in which each of k (and only k) fixed elements of the n-basis occurs exactly s times. From (1.1) and (0.2) we find the exact distribution of $\xi_{nm}(s)$:

$$(1.2)\ \mathbf{P}\{\xi_{nm}(s) = k\} = \binom{n}{k}\binom{n+m-1}{m}^{-1}$$

$$\times \sum_{j=0}^{n-k} (-1)^j \binom{n-k}{j}\binom{n+m-(s+1)(k+j)-1}{m-s(k+j)}$$

for $k = 0, 1, \dots, n$. In particular, taking (0.3) into account, we conclude that the distribution of the random variable $\xi_{nm}(0)$, the number of elements of the n-basis which are absent in a random m-sample, is given by the formula

$$(1.3) \qquad \mathbf{P}\{\xi_{nm}(0) = k\} = \binom{n}{k}\binom{m-1}{n-k-1}\binom{n+m-1}{m}^{-1}$$

for $k = 0, 1, \dots, n$. This distribution coincides with that of the number of empty cells under the random equiprobable allocation of n unlike cells by m identical objects.

Formulae (1.2) are too complicated to be applied in calculations for large n and m and thus it is natural to seek a method for finding the limiting distributions of $\xi_{nm}(s)$ as n and (or) m tend to infinity. This problem remains to be solved in the following theorem.

Theorem 1.1 *Let s be fixed and $n, m \to \infty$ in such a way that $n^2 m^{-1} \to \lambda \in (0, \infty)$. Then*

$$\lim_{n, m \to \infty} \mathbf{P}\{\xi_{nm}(s) = k\} = \frac{\lambda^k}{k!} e^{-\lambda}, \qquad k = 0, 1, \ldots .$$

Proof We consider the probability generating function of $\xi_{nm}(s)$:

$$f_{nm}(x, s) := \sum_{k=0}^{n} \mathbf{P}\{\xi_{nm}(s) = k\} x^k$$

and the generating function of its binomial moments:

$$g_{nm}(x, s) := \sum_{k=0}^{n} B_k(n, m) x^k.$$

Multiplying both sides of equality (0.2) by $\binom{n}{k} x^k$ and summing over k from 0 to n we obtain

$$(1.4) \qquad \sum_{m=0}^{\infty} \binom{n+m-1}{m} f_{nm}(x, s) t^m = \left(\frac{1}{1-t} + (x-1) t^s \right)^n.$$

By the equality $g_{nm}(x, s) = f_{nm}(x+1, s)$, we have

$$\sum_{m=0}^{\infty} \binom{n+m-1}{m} g_{nm}(x, s) t^m = \left(\frac{1}{1-t} + x t^s \right)^n.$$

Calculating the coefficients of $t^m x^k$ in both sides of the preceding equality we see that the kth binomial moment of $\xi_{nm}(s)$ is given by the formula

$$(1.5) \qquad B_k(n, m) = \frac{(n)_k \, (n-1)_k \, (m)_{ks}}{k! \, (n+m-1)_{k(s+1)}}.$$

Hence it follows that

$$\lim_{n, m \to \infty} B_k(n, m) = \frac{\lambda^k}{k!}, \qquad k = 0, 1, \ldots .$$

This proves Theorem 1.1.   □

Setting $s = 0$ in Theorem 1.1 we obtain the following statement.

Corollary *Under the conditions of Theorem 1.1 the limiting distribution of $\xi_{nm}(0)$ is Poisson with parameter λ.*

We now study the asymptotic behavior of $\xi_{nm}(0)$ under other conditions.

Theorem 1.2 *Let* $n, m \to \infty$ *in such a way that*

$$\frac{n}{m+n} \longrightarrow p, \qquad \frac{m}{m+n} \longrightarrow q, \qquad p+q=1, \quad p, q > 0.$$

If $k = (n+m)p^2 + o\big((n+m)^{2/3}\big)$ *then*

$$\mathbf{P}\{\xi_{nm}(0) = k\} = \frac{1}{pq\sqrt{2\pi(n+m)}} \exp\left\{-\frac{(k-(n+m)p^2)^2}{2(n+m)p^2q^2} + o(1)\right\}.$$

Proof Let us consider the random variable

$$\eta_{nm} = \frac{\xi_{nm}(0) - (n+m)p^2}{pq\sqrt{n+m}}.$$

To prove the theorem it suffices to establish that if $w = o\big((n+m)^{1/6}\big)$ then

$$\mathbf{P}\{\eta_{nm} = w\} = \frac{1}{\sqrt{2\pi}}\, e^{-w^2/2}\, dw\,(1 + o(1)), \qquad dw = \frac{1}{pq\sqrt{n+m}}.$$

Using relation (1.3) and Stirling's formula for $n!$ we derive

$$(1.6) \quad \mathbf{P}\{\xi_{nm}(0) = k\} = \frac{1}{\sqrt{2\pi}}\, n^{2n}\, m^{2m}\, (n+m)^{1/2-n-m}$$

$$\times\, (m-n+k)^{-m+n-k-1/2}\, k^{-k-1/2}$$

$$\times\, (n-k)^{2(k-n)}\big(1 + O(m^{-1})\big), \quad n \to \infty.$$

Setting

$$N = m+n, \qquad k = Np^2 + wpq\sqrt{N}$$

and making the necessary simplifications in formula (1.6), we obtain

$$\mathbf{P}\{\eta_{nm} = w\} = \mathbf{P}\{\xi_{nm}(0) = k\} = \frac{1}{pq\sqrt{2\pi N}} \exp\left\{-\frac{w^2}{2} + O\left(\frac{w^3}{\sqrt{N}}\right)\right\},$$

thereby completing the proof of the theorem.   □

3.1.2 Distribution of the number of occurrences
of fixed elements of an n-basis

Let $\varkappa_{nm}(s)$ be the total number of occurrences of s fixed elements in a random m-sample of a commutative nonsymmetric n-basis. The random variable $\varkappa_{nm}(s)$ may be viewed as the total number of objects contained in s cells under the random equiprobable allocation of n unlike cells by m identical objects. From formula (0.5) it is not difficult to find the exact distribution of $\varkappa_{nm}(s)$:

$$\mathbf{P}\{\varkappa_{nm}(s) = k\} = \binom{k+s-1}{k}\binom{n+m-s-k-1}{m-k}\binom{n+m-1}{m}^{-1}$$

for $k = 0, 1, \ldots, m$. We now study the limiting behavior of the distribution of $\varkappa_{nm}(s)$.

Theorem 1.3 *Let $n, m \to \infty$. The following two statements are valid:*
(1) If s is fixed and

$$\frac{n}{m+n} \longrightarrow p, \qquad \frac{m}{m+n} \longrightarrow q, \qquad p+q=1, \ p, \ q > 0,$$

then the distribution of $\varkappa_{nm}(s)$ converges to the Pascal distribution with parameters (s, p):

$$\lim_{n,\,m\to\infty} \mathbf{P}\{\varkappa_{nm}(s) = k\} = \binom{k+s-1}{k} p^s \, q^k, \qquad k = 0, 1, \ldots .$$

(2) If $s \to \infty$ and $s m n^{-1} \to \lambda \in (0, \infty)$ then the distribution of $\varkappa_{nm}(s)$ converges to a Poisson distribution with parameter λ:

$$\lim_{n,\,m\to\infty} \mathbf{P}\{\varkappa_{nm}(s) = k\} = \frac{\lambda^k}{k!} \, e^{-\lambda}, \qquad k = 0, 1, \ldots .$$

Proof Let

$$f_{nm}(x) = \sum_{k=0}^{m} \mathbf{P}\{\varkappa_{nm}(s) = k\} \, x^k,$$

$$g_{nm}(x) = \sum_{k=0}^{m} B_k(n, m) \, x^k$$

be the generating function of $\varkappa_{nm}(s)$ and the generating function of the binomial moments of $\varkappa_{nm}(s)$ respectively. Multiplying both sides of equality (0.4) by $\binom{k+s-1}{k} x^k$ and then summing over k we obtain

$$\sum_{m=0}^{\infty} \binom{n+m-1}{m} f_{nm}(x) \, t^m = (1 - x t)^{-s} \, (1 - t)^{-(n-s)}.$$

Hence it follows that

$$\sum_{m=0}^{\infty} \binom{n+m-1}{m} g_{nm}(x) \, t^m = \left(1 - \frac{x t}{1-t}\right)^{-s} (1 - t)^{-n}.$$

By comparing the coefficients of $t^m x^k$ on both sides of the equality we find the following expressions for the kth binomial moment of $\varkappa_{nm}(s)$:

$$B_k(n; m) = \binom{k+s-1}{k} \frac{(m)_k}{(n+k-1)_k}, \qquad k = 0, 1, \ldots, m.$$

Provided that condition (1) of the theorem holds we have

$$\lim_{n,\,m\to\infty} B_k(n;\,m) = \binom{k+s-1}{k}\left(\frac{q}{p}\right)^k, \qquad k = 0,\,1,\dots.$$

Likewise, under condition (2) we deduce that

$$\lim_{n,\,m\to\infty} B_k(n\,;m) = \frac{\lambda^k}{k!}, \qquad k = 0,\,1,\dots. \qquad \square$$

Denote by $\bar{\varkappa}_{nm}(s)$ the total number of occurrences of s fixed elements of an n-basis in a random m-sample given that each element of the n-basis considered appears at least once. By formula (0.6) the exact distribution of $\bar{\varkappa}_{nm}(s)$ takes the form

$$\mathbf{P}\{\bar{\varkappa}_{nm}(s) = k\} = \binom{k-1}{s-1}\binom{m-k-1}{n-s-1}\binom{m-1}{n-1}^{-1},$$

$$k = s,\, s+1,\,\dots,\,n.$$

Theorem 1.4 *Let $n,\,m \to \infty$. The following two statements are valid:*
(1) *If s is fixed and*

$$n\,m^{-1} \to p, \qquad (m-n)\,m^{-1} \to q, \qquad p+q = 1,\ p,\,q > 0,$$

then

$$\lim_{n,\,m\to\infty} \mathbf{P}\{\bar{\varkappa}_{nm}(s) - s = k\} = \binom{k+s-1}{k}\,p^s\,q^k, \qquad k = 0,\,1,\dots.$$

(2) *If $s \to \infty$ and $s\,(m-n)\,n^{-1} \to \lambda \in (0,\,\infty)$ then*

$$\lim_{n,\,m\to\infty} \mathbf{P}\{\bar{\varkappa}_{nm}(s) - s = k\} = \frac{\lambda^k}{k!}\,e^{-\lambda}, \qquad k = 0,\,1,\dots.$$

Proof By the method used in the proof of Theorem 1.3 one can obtain the following formula for the kth binomial moment of the random variable $\bar{\varkappa}_{nm}(s) - s$:

$$\bar{B}_k(n,\,m) = \binom{k+s-1}{k}\frac{(m-n)_k}{(n+k-1)_k}, \qquad k = 0,\,1,\dots.$$

Now the statements of the theorem follow from the corresponding limiting expressions for the binomial moments of $\bar{\varkappa}_{nm}(s) - s$ under conditions (1) and (2) respectively. $\qquad \square$

3.2 Series in random sequences

3.2.1 The number of series having a prescribed number of 0s and 1s

We consider the uniform distribution on the set of all such sequences of length m, each of which contains m_0 zeros and m_1 units, $m_0 + m_1 = m$, by assigning to each such sequence the probability $\binom{m}{m_0}^{-1}$. Let ξ_0 be the number of series of zeros in a random sequence. From (0.7) we see that

$$(2.1) \quad \mathbf{P}\{\xi_0 = k\} = \binom{m_0 - 1}{k - 1}\binom{m_1 + 1}{k}\binom{m}{m_0}^{-1}, \quad k = 0, 1, \ldots, m_0.$$

The following formula holds for the binomial moments of ξ_0:

$$(2.2) \qquad B_k = \binom{m_0}{k}\frac{(m_1 + 1)_k}{(m)_k}, \qquad k = 0, 1, \ldots, m_0.$$

Using (2.2) it is not difficult to show that the mean and variance of ξ_0 are as follows:

$$\mathbf{E}\,\xi_0 = \frac{m_0(m_1 + 1)}{m}, \qquad \mathrm{Var}\,\xi_0 = \frac{(m_0)_2(m_1 + 1)_2}{m(m)_2}.$$

If $m_0 m^{-1} \to \alpha$ and $m_1 m^{-1} \to \beta$ as $m \to \infty$, where $\alpha + \beta = 1$, $\alpha\beta > 0$, then the mean and variance of ξ_0 have the asymptotic representations

$$(2.3) \qquad \begin{cases} \mathbf{E}\,\xi_0 = \alpha\beta m\,(1 + o(1)), \\ \mathrm{Var}\,\xi_0 = (\alpha\beta)^2\,m\,(1 + o(1)). \end{cases}$$

We study the limiting distribution of ξ_0 under the same conditions.

Theorem 2.1 *If $m_0 m^{-1} \to \alpha$ and $m_1 m^{-1} \to \beta$ as $m \to \infty$, where $\alpha + \beta = 1$, $\alpha\beta > 0$, then for $k = \alpha\beta m + o(m^{2/3})$*

$$\mathbf{P}\{\xi_0 = k\} = \frac{1}{\alpha\beta\sqrt{2\pi m}}\exp\left\{-\frac{1}{2}\left(\frac{k - \alpha\beta m}{\alpha\beta\sqrt{m}}\right)^2 + o(1)\right\}.$$

Proof It suffices to show that if $w = o(m^{1/6})$ then the random variable $\eta_0 = (\xi_0 - \alpha\beta m)(\alpha\beta\sqrt{m})^{-1}$ satisfies the relation

$$\mathbf{P}\{\eta_0 = w\} = \frac{1}{\sqrt{2\pi}}e^{-w^2/2}\,dw\,(1 + o(1)), \qquad dw = \frac{1}{\alpha\beta\sqrt{m}}.$$

Applying Stirling's formula to $n!$ as $n \to \infty$, we derive from (2.1) the expansion

$$(2.4) \qquad \mathbf{P}\{\xi_0 = k\} = \frac{1}{\sqrt{2\pi}}\,(k-1)^{1/2-k}\,(m_0-1)^{m_0-1/2}$$

$$\times\,(m_1+1)^{m_1+3/2}\,k^{-k-1/2}\,m^{-m-1/2}$$

$$\times\,m_0^{m_0+1/2}\,m_1^{m_1+1/2}\,(m_0-k)^{k-m_0-1/2}$$

$$\times\,(m_1-k+1)^{k-m_1-3/2}\left(1+O(m^{-1})\right).$$

Setting $k = \alpha\beta m + w\alpha\beta\sqrt{m}$ and making the necessary simplifications in (2.4) we obtain

$$\mathbf{P}\{\eta_0 = w\} = \mathbf{P}\{\xi_0 = k\} = \frac{1}{\alpha\beta\sqrt{2\pi m}}\,\exp\left\{-\frac{w^2}{2}+O\!\left(\frac{w^3}{\sqrt{N}}\right)\right\}.$$

Hence the theorem follows. $\square$

3.2.2 Series in Bernoulli's scheme

Let a sequence of 0s and 1s be a realization of m independent trials in Bernoulli's scheme with probability of success p and probability of failure q, $p+q=1$. Here we agree to consider a trial as a success if 0 appears and as a failure if 1 appears. Let us denote by $\xi_0(p)$ the number of series of 0s in the sequence of m trials. Using the total probability formula we have

$$\mathbf{P}\{\xi_0(p) = k\} = \sum_{m_0=0}^{m} \mathbf{P}(m_0)\,\mathbf{P}\{\xi_0(p) = k/m_0\},$$

where $\mathbf{P}(m_0)$ is the probability of m_0 successes in m trials. Obviously,

$$\mathbf{P}(m_0) = \binom{m}{m_0} p^{m_0}\,q^{m_1}, \qquad m_0 + m_1 = m.$$

It is also clear that

$$\mathbf{P}\{\xi_0(p) = k/m_0\} = \mathbf{P}\{\xi_0 = k\}.$$

Applying formula (2.1) we finally obtain

$$\mathbf{P}\{\xi_0(p) = k\} = \sum_{m_0=0}^{m} \binom{m_0-1}{k-1}\binom{m_1+1}{k} p^{m_0}\,q^{m_1}, \qquad m_0 + m_1 = m.$$

The exact distribution allows us to derive the explicit expression for the kth binomial moment of $\xi_0(p)$, $k = 0, 1, \ldots, m$:

$$(2.5) \qquad B_k(m; p) = \binom{m-k+1}{k} p^k q^{k-1} - \binom{m-k}{k} p^{k+1} q^{k-1}.$$

Hence it follows that

$$(2.6) \qquad \begin{cases} \mathbf{E}\,\xi_0(p) = mpq + p^2, \\ \operatorname{Var} \xi_0(p) = mpq\,(1 - 3pq) + (p^2 + 2p^2 q - 6p^3 q - p^4). \end{cases}$$

Theorem 2.2 *If $p = p(m)$, $q = q(m)$ and $mpq \to \lambda \in (0, \infty)$ as $m \to \infty$ then*

$$\lim_{m \to \infty} \mathbf{P}\,\{\xi_0(p) = k\} = \frac{\lambda^k}{k!}\, e^{-\lambda}, \qquad\qquad k = 0, 1, \ldots.$$

Proof Under the conditions given above, formula (2.5) implies that

$$\lim_{m \to \infty} B_k(m; p) = \frac{\lambda^k}{k!}, \qquad\qquad k = 0, 1, \ldots,$$

which completes the proof of the theorem. $\qquad\qquad\qquad\qquad\qquad\qquad$ □

We now study the distribution of $\xi_0(p)$ holding p fixed and letting $m \to \infty$.

Theorem 2.3 *For any fixed p the distribution of the random variable*

$$\eta_0(p) = \frac{\xi_0(p) - mpq}{\sqrt{mpq\,(1 - 3pq)}}$$

converges to the standard normal distribution as $m \to \infty$.

Proof Let us consider the generating function of the binomial moments of $\xi_0(p)$:

$$B_m(t) = \sum_{k=0}^{\infty} B_k(m; p)\, t^k.$$

From (2.5) we know that

$$B_m(t) = \sum_{k=0}^{\infty} \binom{m-k+1}{k} p^k q^{k-1}\, t^k - \sum_{k=0}^{\infty} \binom{m-k}{k} p^{k+1} q^{k-1}\, t^k.$$

Hence we deduce that the probability generating function of $\xi_0(p)$ is of the form

$$(2.7) \qquad P_m(t) = \sum_{k=0}^{\infty} \binom{m-k+1}{k} p^k q^{k-1}\, (t-1)^k$$

$$- \sum_{k=0}^{\infty} \binom{m-k}{k} p^{k+1} q^{k-1}\, (t-1)^k.$$

We consider the double generating function

$$P(x,\, t) := \sum_{m=0}^{\infty} P_m(t)\, x^m, \qquad\qquad P_0(t) := 1\,.$$

By relation (2.7) one can show that

$$P(x;\, t) = \frac{1 + px\,(t-1)}{1 - x - pq\,(t-1)\, x^2}\,.$$

Let x_1 and x_2 be the roots of the equation $pq\,(1-t)\, x^2 - x + 1 = 0$. It is not difficult to check that

$$P(x;\, t) = \frac{1}{\sqrt{1 - 4pq\,(1-t)}}\left[\frac{1 - px_1\,(1-t)}{x - x_1} - \frac{1 - px_2\,(1-t)}{x - x_2}\right],$$

and so

$$P_m(t) = \frac{1}{\theta}\left\{\frac{q - p + \theta}{2q}\left(\frac{1+\theta}{2}\right)^{m+1} + \frac{p - q + \theta}{2q}\left(\frac{1-\theta}{2}\right)^{m+1}\right\},$$

where $\theta = \sqrt{1 - 4pq\,(1-t)}$. Using this result and writing the characteristic function of $\eta_0(p)$ as

$$\varphi_m(t) = e^{-it\, M/\sigma}\, P_m(e^{it/\sigma}),$$

with $M = mpq$ and $\sigma^2 = mpq\,(1 - 3pq)$, we establish the asymptotic expansion

$$\log \varphi_m(t) = -\frac{itM}{\sigma} + \frac{itmpq}{\sigma} - \frac{mpq\,(1 - 3pq)}{\sigma^2}\,\frac{t^2}{2} + O\!\left(\frac{m}{\sigma^3}\right),$$

which means that

$$\lim_{m\to\infty} \varphi_m(t) = e^{-t^2/2}$$

for any fixed t.   □

3.2.3 Series on a circle

We consider the set of all sequences of length m containing m_0 zeros and m_1 units, $m_0 + m_1 = m$, that are located on a circle, and assign to each such sequence the probability $\binom{m}{m_0}^{-1}$. Let $2\xi^c$ be the number of series in such a random sequence. Applying formula (0.8) we obtain

$$(2.8)\quad \mathbf{P}\,\{\xi^c = k\} = \frac{m}{k}\binom{m_0 - 1}{k - 1}\binom{m_1 - 1}{k - 1}\binom{m}{m_0}^{-1}, \qquad k = 0,\, 1,\, \dots\,.$$

Hence we can find the binomial moments of the random variable ξ^c:

$$(2.9) \qquad B_k^c = \frac{(m_0)_k \, (m_1)_k}{k! \, (m-1)_k}, \qquad k = 0, 1, \dots .$$

Exploiting this result it is not difficult to show that the mean and variance of ξ^c can be calculated by the formulae

$$\mathbf{E}\,\xi^c = \frac{m_0 \, m_1}{m-1},$$

$$\operatorname{Var} \xi^c = \frac{m_0 \, m_1}{(m-1)^2 \, (m-2)} \, (m_0 \, m_1 - m + 1).$$

Theorem 2.4 *If $m_0 \, m^{-1} \to \alpha$ and $m_1 \, m^{-1} \to \beta$ as $m \to \infty$, where $\alpha + \beta = 1$, $\alpha\beta > 0$, then for $k = \alpha\beta m + O\left(m^{2/3}\right)$*

$$\mathbf{P}\,\{\xi^c = k\} = \frac{1}{\alpha\,\beta\,\sqrt{2\pi m}} \, \exp\left\{-\frac{1}{2} \left(\frac{k - \alpha\beta m}{\alpha\beta\sqrt{m}}\right)^2 + o(1)\right\}.$$

The **proof** of the theorem coincides almost exactly with that of Theorem 2.1. □

3.2.4 Series on a circle generated by realizations of Bernoulli's scheme

We consider Bernoulli's scheme with probability of success p and assume that elements of a realization of length m of this scheme are located on a circle. From (2.8) we find the exact distribution for the number $\xi^c(p)$ of the series of successes (or failures) in a random sequence of length m:

$$(2.10) \qquad \mathbf{P}\,\{\xi^c(p) = k\} = \frac{m}{k} \sum_{m_0=0}^{\infty} \binom{m_0 - 1}{k - 1}\binom{m_1 - 1}{k - 1} p^{m_0} \, q^{m_1}$$

for all $k = 0, 1, \dots$. Using the equality we find the binomial moments of $\xi^c(p)$:

$$(2.11) \qquad B_k^c(m; p) = \frac{m}{k} \binom{m-k-1}{k-1} p^k \, q^k, \qquad k = 0, 1, \dots .$$

From these relations it is easy to deduce that the mean and variance of $\xi^c(p)$ are equal to

$$\mathbf{E}\,\xi^c(p) = mpq, \qquad \operatorname{Var} \xi^c(p) = mpq\,(1 - 3pq).$$

Theorem 2.5 *If $mpq \to \lambda \in (0, \infty)$ as $m \to \infty$ then*

$$\lim_{k \to \infty} \mathbf{P}\,\{\xi^c(p) = k\} = \frac{\lambda^k}{k!} \, e^{-\lambda}, \qquad k = 0, 1, 2, \dots .$$

The validity of the theorem follows directly from (2.11). □

Theorem 2.6 *For any fixed p and $x \in (-\infty, +\infty)$,*

$$\lim_{m \to \infty} \mathbf{P} \left\{ \frac{\xi^c(p) - mpq}{\sqrt{mpq\,(1 - 3pq)}} < x \right\} = \frac{1}{\sqrt{2\pi}} \int_{-\infty}^{x} e^{-u^2/2}\, du .$$

Proof Set

$$P_0^c(t) := 1, \qquad P_m^c(t) := \sum_{k=0}^{\infty} \mathbf{P}\{\xi^c(p) = k\}\, t^k, \quad m \geq 1,$$

and

$$P^c(x; t) := \sum_{m=0}^{\infty} P_m^c(t)\, x^m.$$

Using formula (2.11) we obtain

$$P^c(x; t) = \frac{1 + p\,q\,(t - 1)\,x^2}{1 - x - p\,q\,(t - 1)\,x^2}$$

and therefore

$$P_m^c(t) = \frac{1}{p\,q\,(1 - t)\,(x_1 - x_2)} \left[\frac{x_1 - 2}{x_1^{m+1}} - \frac{x_2 - 2}{x_2^{m+1}} \right], \qquad m = 1, 2, \ldots ,$$

where x_1 and x_2 are the roots of the equation $pq\,(1 - t)\,x^2 - x + 1 = 0$.
Simplifying the expression we deduce the relation

$$(2.12) \quad P_m^c(t) = \left(\frac{1 + \sqrt{1 - 4pq\,(1 - t)}}{2} \right)^m + \left(\frac{1 - \sqrt{1 - 4pq\,(1 - t)}}{2} \right)^m .$$

Write $M = mpq$ and $\sigma^2 = mpq\,(1 - 3pq)$. Then the characteristic function
of $\xi^c(p)$ takes the form

$$\varphi_m^c(t) = e^{-it\,M/\sigma}\, P_m^c(e^{it/\sigma}).$$

Relation (2.12) shows that

$$\lim_{m \to \infty} \varphi_m^c(t) = e^{-t^2/2}$$

for any fixed t. □

3.3 Probability distributions
for a noncommutative nonsymmetric n-basis

We consider the set of all m-samples of a noncommutative nonsymmetric
n-basis and specify on the set the uniform distribution by assigning to each
sample probability m^{-n}. We will investigate various random variables on
the set of elementary events and study their exact and limiting (as n and
m tend to infinity) distributions. It should be noted that the given space
of elementary events is the basis of a class of problems known under the
common name of the classical occupancy problem. One can see this by

observing that with each m-sample in a noncommutative nonsymmetric n-basis is associated the corresponding allocation of m distinct objects into n unlike cells. Paper [65] contains a comprehensive survey of a large number of articles devoted to studying random variables related to the classical occupancy problem. The detailed proofs of the most important asymptotic results in the field are given in [66]. In the present section we list only certain typical statements of an asymptotic nature that deal with a range of problems. The reader may find a more systematic treatment of the problems in the sources cited and the references therein.

3.3.1 Distributions of the number of elements of an n-basis occurring s times in a random m-sample

We denote by $\xi_{nm}(s)$ the number of elements of a noncommutative non-symmetric n-basis which occur exactly s times in a random m-sample. It is clear that

$$\mathbf{P}\{\xi_{nm}(s) = k\} = n^{-m} \binom{n}{k} D_{nm}^{(k)}(s), \qquad k = 0, 1, \ldots, n,$$

where $D_{nm}^{(k)}(s)$ is the number of m-samples in which k fixed elements (and only these k elements) occur exactly s times. Applying formula (0.10) we see that, for all $k = 0, 1, \ldots, n$,

$$(3.1) \qquad \mathbf{P}\{\xi_{nm}(s) = k\} = n^{-m} \binom{n}{k} \sum_{j=0}^{n-k} (-1)^j \binom{n-k}{j}$$

$$\times \frac{(n-k-j)^{m-s(j+k)} \, m_{s(j+k)}}{(s\,!)^{j+k}} \, .$$

Formula (3.1) also gives the distribution of the number of cells containing s objects under the random equiprobable allocation of m distinct objects into s unlike cells. Setting $s = 0$ in (3.1) and recalling (0.11) we find the distribution of the number of empty cells under the random allocation of m distinct objects into n unlike cells:

$$(3.2) \qquad \mathbf{P}\{\xi_{nm}(0) = k\} = n^{-m} \binom{n}{k} \Delta^{n-k} 0^m, \qquad k = 0, 1, \ldots, n\,.$$

Now we proceed to study the asymptotic distribution of $\xi_{nm}(s)$ as $n, m \to \infty$.

Theorem 3.1 *Let* $s \in \{0, 1, \dots\}$ *and* $\gamma \in (-\infty, +\infty)$ *be fixed and let* $m = n\big(s \log m - (s-1) \log n + \gamma\big) + O(1)$ *as* $n \to \infty$. *Then*

$$\lim_{n,\, m \to \infty} \mathbf{P}\left\{\xi_{nm}(s) = k\right\} = \frac{\lambda^k}{k!}\, e^{-\lambda}, \qquad k = 0, 1, \dots,$$

where $\lambda = e^{-\gamma}/s!$.

Proof We consider the probability generating function and the generating function of binomial moments of $\xi_{nm}(s)$:

$$f_{nm}(x) = \sum_{k=0}^{m} \mathbf{P}\left\{\xi = k\right\} x^k,$$

$$g_{nm}(x) = \sum_{k=0}^{m} B_k^{(s)}(n, m)\, x^k.$$

Multiplying both sides of (0.10) by $\binom{n}{k} x^k$, summing over k from 0 to n and performing obvious simplifications we obtain

$$(3.3) \qquad \sum_{m=0}^{\infty} n^m f_{nm}(x)\, \frac{t^m}{m!} = \left(e^t + (x-1)\, \frac{t^s}{s!}\right)^n.$$

Hence we deduce the representation

$$\sum_{m=0}^{\infty} n^m g_{nm}(x)\, \frac{t^m}{m!} = \left(e^t + x\, \frac{t^s}{s!}\right)^n.$$

Comparing the coefficients of $x^m\, t^m / m!$ on both sides of the preceding equality we find the kth binomial moment of $\xi_{nm}(s)$:

$$(3.4) \qquad B_k^{(s)}(n, m) = \frac{(n)_k\, (m)_{sk}\, (n-k)^{m-sk}}{k!\, (s!)^k\, n^m}, \qquad k = 0, 1, \dots.$$

Hence it follows that

$$\lim_{n,\, m \to \infty} B_k^{(s)}(n, m) = \frac{1}{k!}\left(\frac{e^{-\gamma}}{s!}\right)^k, \qquad k = 0, 1, \dots,$$

which completes the proof of the theorem. $\qquad\qquad\qquad\qquad\qquad\square$

Corollary *Let* $\gamma \in (-\infty, +\infty)$ *be fixed and* $m = n(\log n + \gamma) + O(1)$ *as* $n \to \infty$. *Then the limiting distribution of the random variable* $\xi_{nm}(0)$ *is Poisson with parameter* $\lambda = e^{-\gamma}$.

Using the explicit expression (3.4) it is not difficult to find the mean and variance of $\xi_{nm}(s)$:

$$\mathbf{E}\,\xi_{nm}(s) = \binom{m}{s} \frac{(n-1)^{m-s}}{n^{m-1}}\,,$$

$$\mathrm{Var}\,\xi_{nm}(s) = \frac{(n)_2\,(m)_{2s}\,(n-2)^{m-2s}}{(s!)^2\,n^m} + \frac{(m)_s\,(n-1)^{m-s}}{s!\,n^{m-1}}$$

$$- \left[\frac{(m)_s\,(n-1)^{m-s}}{s!\,n^{m-1}}\right]^2.$$

Letting $s = 0$ in (3.4) we obtain the kth binomial moment of $\xi_{nm}(0)$:

$$B_k(n,\,m) = \binom{n}{k}\left(1 - \frac{k}{n}\right)^m,\qquad k = 0,\,1,\,\dots,$$

and, therefore,

$$\mathbf{E}\,\xi_{nm}(0) = n\left(1 - \frac{1}{n}\right)^m,$$

$$\mathrm{Var}\,\xi_{nm}(0) = n\,(n-1)\left(1 - \frac{2}{n}\right)^m + n\left(1 - \frac{1}{n}\right)^m - n^2\left(1 - \frac{1}{n}\right)^{2m}.$$

If $m\,n^{-1} \to a \in (0,\,\infty)$ as $n,\,m \to \infty$ then the mean and variance of $\xi_{nm}(0)$ have the asymptotic representations

$$\begin{cases} \mathbf{E}\,\xi_{nm}(0) = n\,e^{-a}\,(1 + o(1))\,, \\[2mm] \mathrm{Var}\,\xi_{nm}(0) = n\left(e^{-a} - (1+a)\,e^{-2a}\right)(1 + o(1))\,. \end{cases}$$

(3.5)

We now study the asymptotic distribution of $\xi_{nm}(0)$ under the conditions imposed above.

Theorem 3.2 ([92]) *Let $n,\,m \to \infty$ in such a way that $m\,n^{-1} \to a \in (0,\,\infty)$. Then the distribution of the random variable*

$$\eta_{nm} = \frac{\xi_{nm}(0) - n\,e^{-a}}{\sqrt{n\left(e^{-a} - (a+1)\,e^{-2a}\right)}}$$

is asymptotically normal with parameters $(0,1)$ as $n,\,m \to \infty$.

Proof Let $\varphi_{nm}(x)$ be the characteristic function of $\xi_{nm}(0)$. Using (3.3) and Cauchy's formula we obtain

$$\varphi_{nm}(x) = \frac{m!}{2\pi\,i\,n^m} \oint_C \left(e^{ix} - 1 + e^z\right)^n \frac{dz}{z^{m+1}}\,,$$

where C is an arbitrary closed contour in the complex plane z enclosing the origin. Put

$$\theta_1 := e^{-a}, \qquad \theta_2 := \sqrt{e^{-a} - (1+a)\,e^{-2a}}\,, \qquad \theta := \theta_1/\theta_2$$

and take for the contour of integration C the circle of radius a with center at the origin. Then

$$\varphi_{nm}\left(\frac{x}{\theta_2\sqrt{n}}\right)$$

$$= \frac{m!}{2\pi\,(an)^m} \int_{-\pi}^{\pi} \left(\exp\left\{\frac{i\,x}{\theta_2\sqrt{n}}\right\} - 1 + \exp\left\{a\,e^{iu}\right\}\right)^n \exp\left\{-i\,m\,u\right\} du\,.$$

Changing the variable in the integral by the relation $v = u\sqrt{an}$ and applying Stirling's formula to estimate $m!$ we find the following representation for the characteristic function of η_{nm}:

$$\psi_{nm}(x) = \exp\left\{-i\,\theta\,x\sqrt{n}\right\} \varphi_{nm}\left(\frac{x}{\theta_2\sqrt{n}}\right)$$

$$= \frac{1}{\sqrt{2\pi}} \int_{-\pi\sqrt{an}}^{\pi\sqrt{an}} E_n^n \exp\left\{-i\left(v\sqrt{an} + x\,\theta\sqrt{an}\right)\right\} dv \left(1 + O\!\left(\frac{1}{n}\right)\right),$$

where

$$E_n := \left(\exp\left\{\frac{i\,x}{\theta_2\sqrt{n}}\right\} - 1\right) e^{-a} + \exp\left\{a\left(\exp\left\{\frac{i\,v}{\sqrt{an}}\right\} - 1\right)\right\}.$$

We write the integral J entering the expression for $\psi_{nm}(x)$ as a sum of three integrals:

$$J = J_1 + J_2 + J_3 = \int_{-\pi\sqrt{an}}^{-\sqrt[3]{an}} + \int_{-\sqrt[3]{an}}^{\sqrt[3]{an}} + \int_{\sqrt[3]{an}}^{\pi\sqrt{an}}\,.$$

For $-\sqrt[3]{an} \le v \le \sqrt[3]{an}$ and any fixed x we have the asymptotic expansion

$$n \log E_n - i\,v\sqrt{an} - i\,x\,\theta\sqrt{n} = -\frac{x^2}{2} - \frac{[v - \theta\,x\sqrt{a}]^2}{2} + O\!\left(\frac{v^3}{\sqrt{n}}\right).$$

Hence it follows that

$$\lim_{n,\,m\to\infty} \frac{1}{\sqrt{2\pi}}\,J_2 = e^{-x^2/2}\,.$$

One can show that

$$\lim_{n,\,m\to\infty} |J_1| = \lim_{n,\,m\to\infty} |J_3| = 0.$$

Therefore, for any fixed x,

$$\lim_{n,\,m\to\infty} \psi_{nm}(x) = e^{-x^2/2}.\qquad\square$$

For $s \in \{0, 1, 2, \ldots\}$ and $a \in (0, \infty)$ we put

$$\theta_1(s) := \frac{a^s}{s!}\, e^{-a}, \qquad \theta_2(s) := \frac{a^s}{s!}\, e^{-a}\left[1 - \frac{a^s}{s!}\, e^{-a}\left(1 + \frac{(a-s)^2}{a}\right)\right].$$

Employing the method applied in the proof of Theorem 3.2 one can establish the following statement.

Theorem 3.3 *Let $n, m \to \infty$ in such a way that $m n^{-1} \to a \in (0, \infty)$. Then, for any fixed s, the random variable*

$$\eta_{nm}(s) = \frac{\xi_{nm}(s) - \theta_1(s)\, n}{\sqrt{\theta_2(s)\, n}}$$

is asymptotically normal with parameters $(0,1)$ as $n, m \to \infty$.

By considering a scheme of independent allocation of m distinct objects into n unlike cells we derive an equivalent statement from Theorem 3.3.

Corollary *If m distinct objects are distributed at random, equiprobably and independently into n unlike cells and $\xi_{nm}(s)$ is the number of cells containing s objects then, under the conditions of Theorem 3.3, the random variable $\eta_{nm}(s)$ is asymptotically normal with parameters $(0,1)$ as $n, m \to \infty$.*

Since $\xi_{nm}(0)$ is the number of empty cells, a similar statement concerning the asymptotic normality of this random variable takes place. The obvious reformulation of Theorem 3.1 gives the corresponding statement regarding the convergence of the distribution of the number of cells containing s objects to a Poisson distribution.

3.3.2 Sequential construction of m-samples

We consider a process for sequentially constructing m-samples setting $m = 1, 2, \ldots$ and assuming that the corresponding elements of a noncommutative nonsymmetric n-basis are chosen at random and independently. This process corresponds to the sequential allocation of distinct objects into n unlike cells.

So, let us select a sequence of elements of an n-basis at random and independently and let $\eta(s)$ denote the number of distinct elements chosen in the first s steps and $\nu_k := \min\{s: \eta(s) = k\}$. It is clear that ν_k is equal to the minimal number of distinct objects needed to occupy k unlike cells provided that the total number of cells is n and, at each step, the cell in which an object should be placed is selected at random, equiprobably and independently of the preceding steps.

The distributions of the random variables ν_k and $\xi_{nm}(0)$ are related by the obvious equality

$$\mathbf{P}\{\nu_k \leq m\} = \mathbf{P}\{\xi_{nm}(0) \leq n - k\}.$$

Hence it follows that

$$\mathbf{P}\{\nu_k \leq m\} = \frac{1}{n^m} \sum_{j=0}^{n-k} \binom{n}{j} \Delta^{n-j} 0^m.$$

Let us calculate the generating function of ν_k. With this aim, we consider the random variables $\delta_1 = \nu_1$ and $\delta_k = \nu_k - \nu_{k-1}$, $k = 2, 3, \ldots$. It is clear that $\delta_1, \delta_2, \ldots, \delta_k$ are independent and

$$\mathbf{P}\{\delta_k = j\} = \left(\frac{k-1}{n}\right)^{j-1} \left(1 - \frac{k-1}{n}\right), \qquad j = 1, 2, \ldots.$$

Hence we find the generating functions of δ_k and ν_k:

$$(3.6) \qquad g_{kn}(x) = \left(1 - \frac{k-1}{n}\right) x \left(1 - \frac{k-1}{n} x\right)^{-1},$$

$$(3.7) \qquad f_{kn}(x) = \prod_{j=1}^{k} g_{jn}(x).$$

By differentiating $g_{kn}(x)$ at the point $x = 1$ we obtain

$$\mathbf{E}\,\delta_k = g'_{kn}(1) = \frac{n}{n-k+1},$$

$$\mathrm{Var}\,\delta_k = g''_{kn}(1) + g'_{kn}(1) - (g'_{kn}(1))^2 = \frac{n(k-1)}{(n-k+1)^2}.$$

Therefore, the mean and variance of ν_k are equal to

$$(3.8) \qquad \begin{cases} \mathbf{E}\,\nu_k = \displaystyle\sum_{j=1}^{k} \frac{n}{n-j+1}, \\[3mm] \mathrm{Var}\,\nu_k = \displaystyle\sum_{j=1}^{k} \frac{n(j-1)}{(n-j+1)^2}. \end{cases}$$

To find asymptotic estimates of $\mathbf{E}\nu_k$ and $\operatorname{Var}\nu_k$ we make use of the following formulae as $n \to \infty$:

$$\sum_{j=1}^{n} \frac{1}{j} = \log n + C + O\left(\frac{1}{n}\right),$$

$$\sum_{j=1}^{n} \frac{1}{j^d} = \frac{n^{1-d}}{1-d} + \zeta(d) + O\left(\frac{1}{n^d}\right),$$

where $C = 0.5772...$ is Euler's constant, $d > 1$ is a real number and $\zeta(z)$ is Riemann's ζ-function. In particular, the first relation gives, for the mean number of trials until all the elements of the n-basis occur, the asymptotic representation

$$\mathbf{E}\nu_n = n\left(\log n + C\right) + O(1), \qquad n \to \infty.$$

Letting $n, k \to \infty$ and assuming $k = o(n)$ and $n^{1/2}k^{-1} \to 0$ we obtain from (3.8)

$$\mathbf{E}\nu_k = n \log \frac{n}{n-k} + O(1),$$

$$\operatorname{Var}\nu_k = \frac{nk}{n-k} - n \log \frac{n}{n-k} + O(1).$$

Under the same conditions the third absolute central moment of δ_j is evaluated as

(3.9) $$\mathbf{E}\left|\delta_j - \frac{n}{n-j+1}\right|^3 = O\left(\frac{j}{n}\right),$$

and the estimate is uniform for $j = 1, 2, \dots, k$.

We introduce the notations

$$M_1 := n \log \frac{n}{n-k}, \qquad \sigma^2 := \frac{nk}{n-k} - n \log \frac{n}{n-k}.$$

Theorem 3.4 ([66, 92]) *Let $k, n \to \infty$. The following two statements are valid:*

(a) *If $k = o(n)$ and $k^{-1}n^{1/2} \to 0$ then the limiting distribution of the random variable $(\nu_k - M_1)\sigma^{-1}$ is normal with parameters $(0,1)$.*

(b) *If $\frac{1}{2}k^2 n^{-1} \to \lambda \in (0, \infty)$ then the limiting distribution of the random variable $\nu_k - k$ is Poisson with parameter λ.*

Proof To prove (a) we apply Lyapunov's theorem, keeping in mind that $\nu_k = \delta_1 + \delta_2 + \cdots + \delta_k$, where $\delta_1, \delta_2, \dots, \delta_k$ are independent random variables. We have

$$B_k^2 = \operatorname{Var}\nu_k = \frac{nk}{n-k} - n \log \frac{n}{n-k} + O(1) > C_1 \frac{k^2}{n},$$

where C_1 is an absolute constant. Estimate (3.9) gives

$$C_k^3 = \sum_{j=1}^{k} \mathbf{E} \left| \delta_j - \frac{n}{n-j+1} \right|^3 = O\left(\frac{k^2}{n}\right).$$

Therefore, if $k, n \to \infty$ and $k^{-1} n^{1/2} \to 0$ then

$$\frac{C_k}{B_k} = O\left(\left(\frac{\sqrt{n}}{k} \right)^{1/3} \right) \longrightarrow 0.$$

Thus, all the conditions of Lyapunov's theorem hold and item (a) is proved.

It follows from (3.7) that the generating function $\bar{f}_{kn}(x)$ of the random variable $\nu_k - k$ is of the form

$$\bar{f}_{kn}(x) = \prod_{j=1}^{k} \left[\left(1 - \frac{j-1}{n} \right) x \left(1 - \frac{j-1}{n} x \right)^{-1} \right].$$

If the conditions of item (b) hold then

$$\bar{f}_{kn}(x) = (x-1) \frac{k^2}{2n} + O\left(\frac{k}{n}\right).$$

Hence it follows that

$$\lim_{n \to \infty} \bar{f}_{kn}(x) = e^{\lambda (x-1)}$$

for any fixed $x \in [0, 1)$. □

3.3.3   Distribution of the number of occurrences of a fixed element of an n-basis

We denote by $\varkappa_{nm}$ the number of occurrences of a fixed element of a noncommutative nonsymmetric n-basis in a random m-sample. It is clear that

$$\mathbf{P}\{\varkappa_{nm} = k\} = \frac{Q(n, m, k)}{n^m},$$

where $Q(n, m, k)$ is the number of m-samples in which the fixed element occurs k times. It follows from (0.12) that

$$\mathbf{P}\{\varkappa_{nm} = k\} = \binom{m}{k} \frac{1}{n^k} \left(1 - \frac{1}{n} \right)^{m-k}, \qquad k = 0, 1, \ldots, m.$$

The distribution is binomial with m trials and probability of success n^{-1}.

The binomial moments of the distribution are of the form

$$B_k(n, m) = n^{-k} \binom{m}{k}, \qquad k = 0, 1, \ldots, m.$$

The mean and variance of $\varkappa_{nm}(s)$ are

$$\mathbf{E} \, \varkappa_{nm}(s) = \frac{m}{n}, \qquad \text{Var} \, \varkappa_{nm}(s) = \frac{m(n-1)}{n^2}.$$

Using well-known properties of binomial distribution it is not difficult to give a discription of the limiting distributions of $\varkappa_{nm}$.

Theorem 3.5 *Let $n, m \to \infty$. The following two assertions are valid:*

(a) *If $mn^{-1} \to \lambda \in (0, \infty)$ then the limiting distribution of $\varkappa_{nm}$ is Poisson with parameter λ.*

(b) *If $mn^{-1} \to \infty$ then the distribution of the random variable*

$$(\varkappa_{nm} - M)\sigma^{-1},$$

where

$$M = mn^{-1} \quad \text{and} \quad \sigma^2 = \frac{m}{n}\left(1 - \frac{1}{n}\right),$$

is asymptotically normal with parameters $(0, 1)$.

3.3.4 Labeled elements of an n-basis

We fix d elements of a noncommutative nonsymmetric n-basis, $1 \le d \le n$, and call them *labeled* elements. Denote by $\xi_{nm}^{(d)}$ the number of labeled elements not presented in a random m-sample. If $P(j, d)$ is the probability of the occurrence of exactly j labeled elements in a random m-sample and $P(k; j, d)$ is the probability of the absence of exactly k elements of the d-basis in a random j-sample then the total probability formula gives

$$(3.10) \qquad \mathbf{P}\{\xi_{nm}^{(d)} = k\} = \sum_{j=1}^{m} P(j, d)\, P(k; j, d).$$

The following equalities are obvious:

$$(3.11) \qquad P(j, d) = \binom{m}{j}\left(\frac{d}{n}\right)^j \left(1 - \frac{d}{n}\right)^{m-j},$$

$$(3.12) \qquad P(k; j, d) = \binom{d}{k}\frac{\Delta^{d-k}\, 0^j}{d^j}.$$

Substituting (3.11) and (3.12) into (3.10) we have

$$(3.13) \quad \mathbf{P}\{\xi_{nm}^{(d)} = k\} = \binom{d}{k}\frac{\Delta^{d-k}(n-d)^m}{n^m}, \qquad\qquad k = 0, 1, \ldots, d.$$

Hence we find the explicit expression for the kth binomial moment of the random variable $\xi_{nm}^{(d)}$:

$$B_k^{(d)}(n, m) = n^{-m}\binom{d}{k}\sum_{j=k}^{d}\binom{d-k}{j-k}\Delta^{d-j}(n-d)^m.$$

Changing the index of summation j by $j + k$ and observing that

$$\sum_{j=0}^{d-k} \binom{d-k}{j} \Delta^{d-k-j} (n-d)^m = (n-k)^m,$$

we finally obtain

$$(3.14) \qquad B_k^{(d)}(n, m) = \binom{d}{k} \left(1 - \frac{k}{n}\right)^m, \qquad k = 0, 1, \dots, d.$$

In particular, the mean of $\xi_{nm}^{(d)}$ is

$$\mathbf{E}\, \xi_{nm}^{(d)} = d \left(1 - \frac{1}{n}\right)^m.$$

Theorem 3.6 *Let $n, m \to \infty$. The following three statements are valid:*

(a) *If d is fixed and $mn^{-1} \to a \in (0, \infty)$ then*

$$\lim_{n,\, m \to \infty} \mathbf{P}\{\xi_{nm}^{(d)} = j\} = \binom{d}{j} p^j (1-p)^{d-j}, \qquad j = 0, 1, \dots, d,$$

where $p = e^{-a}$.

(b) *If $d \to \infty$ and $mn^{-1} = \log d - \log \lambda + o(1)$, where $\lambda > 0$ is a constant, then*

$$\lim_{n,\, m \to \infty} \mathbf{P}\{\xi_{nm}^{(d)} = j\} = \frac{\lambda^j}{j!}\, e^{-\lambda}, \qquad j = 1, 2, \dots.$$

(c) *If $d \to \infty$, $mn^{-1} \to a \in (0, \infty)$ and $dn^{-1} \to \gamma \in (0, \infty)$ then the random variable*

$$\frac{\xi_{nm}^{(d)} - n\gamma e^{-a}}{\sqrt{n\gamma\left[e^{-a} - (1 + a\gamma) e^{-2a}\right]}}$$

is asymptotically normal with parameters $(0, 1)$.

Proof Because of (3.14) the probability generating function of the random variable $\xi_{nm}^{(d)}$ is of the form

$$f_{nm}^{(d)}(x) = \sum_{k=0}^{d} \binom{d}{k} \left(1 - \frac{k}{n}\right)^m (x - 1)^k.$$

Under the conditions of item (a) the equality

$$\lim_{n,\, m\to\infty} f_{nm}^{(d)}(x) = \left(1 + \frac{x-1}{e^a}\right)^d$$

is valid for any $x \in [0, 1)$. The right-hand side of the relation is the generating function of the number of successes in d trials with probability of success $p = e^{-a}$. Now statement (a) follows from the continuity theorem for generating functions.

There is no difficulty in deducing from (3.14) that, provided the conditions of item (b) hold,

$$\lim_{n,\, m\to\infty} B_k^{(d)}(n, m) = \frac{\lambda^k}{k!}\,, \qquad k = 0, 1, \dots .$$

This proves assertion (b). The proof of statement (c) is similar to that of Theorem 3.2. □

3.4 Matching problem

A partition of the set $\mathbf{X} = \{1, 2, \dots, m\}$ is said to be *matching* if each of its blocks contains at most two elements. To obtain a more explicit view on matchings we are interested in a simplified model of a telephone system with m subscribers labeled by $1, 2, \dots, m$. With each state of the system we associate a matching as follows. Consider elements i and j of the set $\mathbf{X}$ as comprising a block of size 2 if the subscribers i and j hold a conversation. If some subscriber, say k, holds no conversation, we regard the element k as comprising a block of size 1. If Q_m is the number of matchings of an m-element set then, putting $\mathbf{A} = \{1, 2\}$ in (0.21), we obtain

(4.1)
$$\sum_{m=0}^{\infty} Q_m \frac{t^m}{m!} = \exp\left\{t + \tfrac{1}{2}t^2\right\}.$$

Hence we see that

$$Q_m = \sum_{k=0}^{[m/2]} \frac{m!}{k!\,(m-2k)!\,2^k}\,, \qquad Q_0 = 1.$$

Let us denote by Q_{mk} the number of matchings with k blocks of size 2. From (0.23) putting $\mathbf{A} = \{1, 2\}$ we know that

(4.2)
$$\sum_{m=0}^{\infty} \sum_{k=0}^{[m/2]} Q_{mk} \frac{t^m}{m!} x^k = \sum_{m=0}^{\infty} Q_m(x) \frac{t^m}{m!}$$
$$= \exp\left\{t + \tfrac{1}{2}x\,t^2\right\},$$

where

$$Q_m(x) := \sum_{k=0}^{[m/2]} Q_{mk}\, x^k.$$

Hence it follows that

$$Q_{mk} = \frac{m!}{k!\,(m-2k)!\,2^k}\,, \qquad k = 0,\,1,\,\ldots,\,[m/2].$$

We consider the uniform distribution on the set of all matchings of an m-element set. If ξ is the number of blocks of size 2 in a random matching then

$$\mathbf{P}\{\xi = k\} = Q_{mk}\, Q_m^{-1}, \qquad k = 0,\,1,\,\ldots,\,[m/2].$$

The mean number of blocks of size 2 is equal to

(4.3) $\qquad \mathbf{E}\,\xi = \binom{m}{2} Q_{m-2}\, Q_m^{-1}, \qquad k = 2,\,3,\,\ldots\,.$

We will study the asymptotic behavior of ξ as $m \to \infty$. In preparation for this, we introduce an auxiliary lemma.

Lemma 1 *There exists a neighborhood $W = [1-\gamma,\, 1+\gamma]$, $\gamma > 0$, of the point $x = 1$ such that, as $m \to \infty$,*

(4.4) $\qquad Q_m(x) = \dfrac{m!\, e^{(m+R)/2}}{R^m \sqrt{2\pi\,(2m-R)}} \left(1 + o(1)\right), \qquad x \in W,$

where $R = R(x)$ is the unique positive solution of the equation

(4.5) $\qquad x\,R^2 + R = m$

and $o(1) \to 0$ uniformly for $x \in W$.

Proof Using equality (4.2) and Cauchy's formula we have

$$Q_m(x) = \frac{m!}{2\pi i} \oint_C \exp\left\{ z + \tfrac{1}{2} x z^2 \right\} \frac{dz}{z^{m+1}},$$

where C is an arbitrary closed contour in the complex plane enclosing the origin. Taking the circle of radius $R = R(x)$ with center at the origin as the countour C and substituting $z = R\,e^{i\theta}$, $-\pi \le \theta \le \pi$, into the integral, we obtain

(4.6) $\qquad Q_m(x) = \dfrac{m!\, \exp\left\{ \tfrac{1}{2} x R^2 + R \right\}}{2\pi\, R^m} \times J,$

where

$$J := \int_{-\pi}^{\pi} \exp\{f(\theta, R, x)\}\, d\theta$$

and

$$f(\theta, R, x) := \tfrac{1}{2} x R^2 (e^{2i\theta} - 1) + R(e^{i\theta} - 1) - i m \theta.$$

We take $\varepsilon = R^{-3/4}$ and represent the integral J as follows:

(4.7) $$J = J_1 + J_2 + J_3 = \int_{-\pi}^{-\varepsilon} + \int_{-\varepsilon}^{\varepsilon} + \int_{\varepsilon}^{\pi}.$$

For $-\pi \le \theta \le -\varepsilon$ and $\varepsilon \le \theta \le \pi$ the estimate

$$\mathrm{Re}\, f(\theta, R, x) = -C\, R^{1/2} + O(R^{-1/2})$$

is true, where $\mathrm{Re}\, z$ is the real part of z and C is a positive constant. Hence we deduce

(4.8) $$|J_1| = O\!\left(e^{-C_1\sqrt{R}}\right), \qquad |J_3| = O\!\left(e^{-C_2\sqrt{R}}\right),$$

where C_1 and C_2 are positive constants.

We consider the derivatives of the function $f(\theta, R, x)$ with respect to θ at $\theta = 0$:

$$f'(0, R, x) = i\,[\,x R^2 + R - m\,],$$

$$f^{(k)}(0, R, x) = i^k\,[\,x\, 2^{k-1} R^2 + R\,], \qquad k = 2, 3, \ldots\,.$$

We take $R = R(x)$ to be the maximal solution of equation (4.5) and expand $f(\theta, R, x)$ in a Maclaurin series in the neighborhood $-\varepsilon \le \theta \le \varepsilon$:

$$f(\theta, R, x) = -[\,2 x R^2 + R\,]\,\frac{\theta^2}{2} + \Delta, \qquad \Delta := \sum_{k=3}^{\infty} i^k\,[\,2^{k-1} x\, R^2 + R\,]\,\frac{\theta^k}{k!}.$$

It is not difficult to show that

$$|\Delta| = O(R^{-1/4}).$$

Using this estimate we find that

$$J_2 = \int_{-\varepsilon}^{\varepsilon} \exp\left\{-\tfrac{1}{2}\theta^2\left(2xR^2+R\right)\right\}\, d\theta \left(1+o(1)\right)$$

uniformly for $x \in W$. Put $\varphi^2 = (2xR^2+R)\,\theta^2$. Then

$$J_2 = \frac{1}{\sqrt{2xR^2+R}} \int_{-\delta}^{\delta} e^{-\varphi^2/2}\, d\varphi \left(1+o(1)\right),$$

where $\delta \geq C_3 R^{1/4}$ and C_3 is a positive constant. Thus, the asymptotic formula

$$(4.9) \qquad J_2 = \sqrt{\frac{2\pi}{2xR^2+R}}\left(1+o(1)\right)$$

is valid uniformly for $x \in W$.

Taking into account exponential estimates (4.8) and combining (4.6), (4.7) and (4.9) we obtain

$$Q_m(x) = \frac{m!\,\exp\left\{\tfrac{1}{2}xR^2+R\right\}}{2\pi R^m}\sqrt{\frac{2\pi}{2xR^2+R}}\left(1+o(1)\right).$$

Hence, performing the necessary simplifications by equation (4.5), we conclude that Lemma 1 is valid. $\square$

Corollary 1 *The number Q_m of matchings of an m-element set is given, as $m \to \infty$, by the asymptotic formula*

$$(4.10) \qquad Q_m = \frac{m^{m/2}}{\sqrt{2}}\,\exp\left\{-\tfrac{1}{2}m+\sqrt{m}-\tfrac{1}{4}\right\}\left(1+o(1)\right).$$

Proof Applying Stirling's formula to $m!$ as $m \to \infty$ we derive from Lemma 1 that

$$(4.11) \qquad Q_m = Q_m(1) = \left(\frac{m}{r}\right)^m e^{-(m-r)/2}\sqrt{\frac{m}{2m-r}}\left(1+o(1)\right),$$

where r is the maximal solution of the equation $r^2+r=m$.

One can deduce the following asymptotic representation for R from equation (4.5):

$$(4.12) \qquad R = \sqrt{\frac{m}{x}}\left(1-\frac{1}{2\sqrt{xm}}+\frac{1}{8xm}+O\left(\frac{1}{m^{3/2}}\right)\right),$$

where the remainder term tends to 0 uniformly for $x \in W$. Hence, setting $x=1$, we obtain

$$(4.13) \qquad r = \sqrt{m}\left(1-\frac{1}{2\sqrt{m}}+\frac{1}{8m}+O\left(\frac{1}{m^{3/2}}\right)\right).$$

The preceding relation implies

$$r^m = m^{m/2}e^{-\sqrt{m}/2}\left(1+o(1)\right).$$

Substituting the estimates for r and r^m into (4.11) we prove (4.10). $\square$

Corollary 2 *If $m \to \infty$ then*

(4.14) $$\mathbf{E}\xi = \tfrac{1}{2} m \left(1 + o(1)\right).$$

Theorem 4.1 *As $m \to \infty$, the distribution of the random variable*

$$\eta = 2m^{-1/4} \left(\xi - \frac{m - m^{1/2}}{2} \right)$$

is asymptotically normal with parameters $(0, 1)$.

Proof According to Lemma 1 and Corollary 1 the generating function of ξ admits, in a neighborhood W of the point $x = 1$, the following asymptotic representation as $m \to \infty$:

$$f_m(x) = \frac{Q_m(x)}{Q_m} = \left(\frac{r}{R} \right)^m e^{(R-r)/2} \left(\frac{2m - r}{2m - R} \right)^{1/2} \left(1 + o(1)\right),$$

where $o(1) \to 0$ uniformly for $x \in W$. By applying estimates (4.12) and (4.13), as $m \to \infty$, we obtain

$$m \, \log \frac{r}{R} = \frac{m}{2} \, \log x + \frac{\sqrt{m}}{2} \left(\frac{1}{\sqrt{x}} - 1 \right) + O\left(\frac{1}{\sqrt{m}} \right),$$

$$\frac{R - r}{2} = \frac{\sqrt{m}}{2} \left(\frac{1}{\sqrt{x}} - 1 \right) - \frac{1}{4} \left(\frac{1}{x} - 1 \right) + O\left(\frac{1}{\sqrt{m}} \right).$$

Therefore,

$$f_m(x) = x^{m/2} \exp\left\{ \sqrt{m} \left(\frac{1}{\sqrt{x}} - 1 \right) - \frac{1}{4} \left(\frac{1}{x} - 1 \right) \right\} \left(1 + o(1)\right)$$

uniformly for $x \in W$. If $\varphi_m(t)$ is the moment generating function of ξ then $\varphi_m(t) = f_m(e^t)$ and, for any t from a neighborhood of the point $t = 0$,

$$\varphi_m\left(\frac{t}{\sigma} \right) = \exp\left\{ \frac{m - \sqrt{m}}{2} \frac{t}{\sigma} + \frac{\sqrt{m}}{4\sigma^2} \frac{t^2}{2} + O\left(\frac{\sqrt{m}}{\sigma^3} \right) \right\} \left(1 + o(1)\right).$$

Take $\sigma^2 = \frac{1}{4} m^{1/2}$. Then

$$\lim_{m \to \infty} \exp\left\{ \frac{-m + \sqrt{m}}{2} \frac{t}{\sigma} \right\} \varphi_m\left(\frac{t}{\sigma} \right) = e^{t^2/2}$$

for any fixed t. Thus, the moment generating function of the random variable η converges, as $m \to \infty$, to $e^{t^2/2}$, being the moment generating function of the standard normal distribution. Now Theorem 4.1 follows from Curtiss' theorem. □

We note in conclusion that the scheme for proving Theorem 4.3 completely coincides with that for proving Theorem 4.3 of Chapter 1, although the double generating function in the first theorem depends on x in a slightly different way than does that in the second theorem.

CHAPTER 4

Random Partitions of Sets

4.0 Introduction

The study of random partitions of sets is a relatively new trend in combinatorics. The results included in this chapter lead readers to the contemporary state of investigation in the given field and create the necessary background for further investigations.

The chapter includes an example of the application of random partition theory to bicycle racing. In giving it we wish only to illustrate the possibilities of the theory rather than to take into account all the sporting and engineering aspects of the problem. The basic combinatorial notions and identities related to partitions used throughout this chapter are given in subsection 3 in the Introduction to Chapter 3.

4.1 Distribution of the number of subsets in a random partition

We consider the uniform distribution on the set of all partitions of an m-element set $\mathbf{X}$ and assign to each such partition the probability T_m^{-1}, where T_m is the mth Bell number. Let ξ_m be the number of subsets (blocks) in a random partition. By equality (0.18) of Chapter 3,

$$\mathbf{P}\left\{\xi_m = k\right\} = \frac{\sigma(m, k)}{T_m}, \qquad k = 0, 1, \ldots, m,$$

where $\sigma(m, k)$ are Stirling's numbers of the second kind.

We consider the generating function

(1.1) $$F_m(x) = \sum_{j=0}^{m} \sigma(m, j)\, x^j.$$

Obviously, $F_m(1) = T_m$ and the generating function of ξ_m takes the form

$$(1.2) \qquad\qquad P_m(x) = \frac{F_m(x)}{F_m(1)}, \qquad\qquad m = 0, 1, \ldots .$$

Observe that representation (0.18) of Chapter 3 can be rewritten as

$$(1.3) \qquad\qquad \sum_{m=0}^{\infty} F_m(x) \frac{t^m}{m!} = \exp\{x(e^t - 1)\}.$$

The kth binomial moment of ξ_m is

$$B_k(m) = \frac{1}{k!} P_m^{(k)}(1),$$

where $P_m^{(k)}(x)$ is the kth derivative of $P_m(x)$. If $F_m^{(k)}(x)$ denotes the kth derivative of $F_m(x)$ then equality (1.3) implies that

$$\sum_{m=0}^{\infty} F_m^{(k)}(x) \frac{t^m}{m!} = (e^t - 1)^k \exp\{x(e^t - 1)\}.$$

Hence it follows that

$$\frac{1}{k!} F_m^{(k)}(1) = \sum_{j=k}^{m} \binom{m}{j} \sigma(j, k) T_{m-j}.$$

From this equality we finally establish

$$B_k(m) = \frac{1}{T_m} \sum_{j=k}^{m} \binom{m}{j} \sigma(j, k) T_{m-j}, \qquad\qquad k = 0, 1, \ldots, m.$$

Setting $k = 1$ in the identity and recalling relation (0.20) of Chapter 3 we find the mathematical expectation of ξ_m:

$$(1.4) \qquad\qquad \mathbf{E}\,\xi_m = \frac{T_{m+1}}{T_m} - 1.$$

By using the asymptotic expansion of Bell's numbers given by formula (0.25) of Chapter 3, we obtain

$$\mathbf{E}\,\xi_m = \frac{m}{r}\,(1 + o(1)), \qquad\qquad m \to \infty,$$

where $r\,e^r = m$. Taking into account representation (0.26) of Chapter 3, we finally conclude that

$$\mathbf{E}\,\xi_m = \frac{m}{\log m}\,(1 + o(1)).$$

One can calculate the variance of ξ_n by the formula for binomial moments. However, we employ another method. By the relations

(1.5) $$\sigma(m, j) = \sigma(m - 1, j - 1) + j\, \sigma(m - 1, j)$$

and

$$\frac{1}{T_m} \sum_{j=1}^{m} j\, \sigma(m, j) = \frac{T_{m+1}}{T_m} - 1,$$

we find that

$$\frac{1}{T_m} \sum_{j=1}^{m} j^2 \sigma(m, j) = \frac{T_{m+2}}{T_m} - 2\, \frac{T_{m+1}}{T_m}.$$

Hence, applying (1.4), we derive

(1.6) $$\operatorname{Var} \xi_m = \frac{T_{m+2}}{T_m} - \left(\frac{T_{m+1}}{T_m} \right)^2 - 1.$$

Using the asymptotic representation of Bell's numbers given by formula (0.25) of Chapter 3, we establish that, as $m \to \infty$,

$$\operatorname{Var} \xi_m = \frac{m}{r^2} \left(1 + o(1) \right),$$

where $r\, e^r = m$. Hence it follows that

(1.7) $$\operatorname{Var} \xi_m = \frac{m}{(\log m)^2} \left(1 + o(1) \right).$$

We now find the limiting distribution of the random variable ξ_m as $m \to \infty$.

Theorem 1.1 *The distribution of the random variable*

$$\eta_m = \frac{\xi_m - \mathbf{E}\, \xi_m}{\sqrt{\operatorname{Var} \xi_m}}$$

converges to the standard normal distribution as $m \to \infty$, that is,

$$\lim_{m \to \infty} \mathbf{P}\{\eta_m < x\} = \frac{1}{\sqrt{2\pi}} \int_{-\infty}^{x} e^{-u^2/2}\, du.$$

To prove the theorem we apply a method for constructing an equivalent scheme of summation of independent random variables [52]. First we establish a lemma concerning the roots of the polynomials $F_m(x)$.

Lemma 1 *For any $m \geq 1$ the roots of the polynomial $F_m(x)$ are distinct, real and nonpositive.*

Proof We prove the lemma by induction. It is easy to show that $F_0(x) = 1$, $F_1(x) = x$ and $F_2(x) = x\,(x+1)$ and thus the initial conditions for applying induction are fulfilled. We now assume that the lemma is valid for all the polynomials $F_n(x)$, $n \leq m-1$. Differentiating $F_m(x)$ in x and using (1.5), we obtain

$$(1.8) \qquad F_m(x) = x\left[F_{m-1}(x) + \frac{d}{dx}\,F_{m-1}(x) \right].$$

By the induction hypothesis, $F_{m-1}(x)$ has $m-1$ distinct, real, nonpositive roots. In the domain $x \in (-\infty, +\infty)$ the function $H_m(x) = F_m(x)\,e^x$ has the same zeros as the polynomial $F_m(x)$. From (1.8) we deduce that

$$F_m(x)\,e^x = x\left[F_{m-1}(x)\,e^x + \left(\frac{d}{dx}\,F_{m-1}(x) \right) e^x \right]$$

and thus

$$(1.9) \qquad H_m(x) = x\,\frac{d}{dx}\,H_{m-1}(x).$$

Observe that

$$\lim_{x \to -\infty} H_{m-1}(x) = \lim_{x \to -\infty} e^x\,F_{m-1}(x) = 0.$$

Therefore, the function $H_{m-1}(x)$ has m roots in the domain $x \in [-\infty, 0)$. By Rolle's theorem each interval on the real line whose endpoints are zeros of the function $H_{m-1}(x)$ contains a point at which

$$\frac{d}{dx}\,H_{m-1}(x) = 0.$$

Consequently, $\frac{d}{dx}H_{m-1}(x)$ has m zeros on the real line (one of them at $x = -\infty$.) Now equality (1.9) implies that $H_m(x)$ is equal to zero at $m+1$ points of the nonpositive part of the real line (one of them at $x = -\infty$). Therefore, the polynomial $F_m(x)$ has m real, nonpositive zeros. $\qquad \square$

Proof of Theorem 1.1 Let $-\alpha_1, -\alpha_2, \dots, -\alpha_{m-1}$ be the roots of the equation $F_m(x) = 0$ other than zero. Then

$$F_m(x) = x\,(x + \alpha_1)\,(x + \alpha_2)\,\cdots\,(x + \alpha_{m-1})$$

and formula (1.2) gives

$$(1.10) \quad P_m(x) = x\left(\frac{x}{1+\alpha_1} + \frac{\alpha_1}{1+\alpha_1} \right) \cdots \left(\frac{x}{1+\alpha_{m-1}} + \frac{\alpha_{m-1}}{1+\alpha_{m-1}} \right).$$

We consider independent random variables $\xi_{m1}, \xi_{m2}, \ldots, \xi_{m,m-1}$, taking the values 0 and 1, such that

$$\mathbf{P}\{\xi_{mi} = 1\} = (1+\alpha_i)^{-1}, \qquad i = 1, 2, \ldots, m-1.$$

If $P_{mi}(x)$ is the generating function of ξ_{mi} then

$$P_{mi}(x) = \frac{x}{1+\alpha_i} + \frac{\alpha_i}{1+\alpha_i}, \qquad i = 1, 2, \ldots, m-1.$$

Equality (1.10) means that the random variable ξ_m can be represented as a sum of independent random variables:

$$\xi_m = \xi_{m1} + \xi_{m2} + \cdots + \xi_{m,m-1} + 1.$$

The mean and variance of ξ_{mi} are

$$\mathbf{E}\,\xi_{mi} = \frac{1}{1+\alpha_i}, \qquad \mathrm{Var}\,\xi_{mi} = \frac{\alpha_i}{(1+\alpha_i)^2}.$$

Let us consider the random variables

$$\eta_{mi} = \frac{\xi_{mi} - \mathbf{E}\,\xi_{mi}}{\sqrt{\mathrm{Var}\,\xi_m}}, \qquad i = 1, 2, \ldots, m-1.$$

It is evident that

$$\eta_m = \eta_{m1} + \eta_{m2} + \cdots + \eta_{m,m-1}$$

and

$$(1.11) \qquad \sum_{i=1}^{m-1} \mathbf{E}\,\eta_{mi} = 0, \qquad \sum_{i=1}^{m-1} \mathrm{Var}\,\eta_{mi} = 1.$$

Observe that the sequence of mutually independent random variables $\xi_{m1}, \xi_{m2}, \ldots, \xi_{m,m-1}$ is a Poisson sequence with $p_k = (1+\alpha_k)^{-1}$, $k = 1, 2, \ldots, m-1$. By virtue of a corollary to Lyapunov's theorem (see Section 2, Chapter 1), the corresponding sequence of random variables $\{\eta_m\}$ is asymptotically normal with parameters $(0, 1)$ if

$$\mathrm{Var}\,\xi_m = \sum_{i=1}^{m-1} p_k\, q_k = \sum_{i=1}^{m-1} \frac{\alpha_i}{(1+\alpha_i)^2} \longrightarrow \infty,$$

which is the case in light of (1.7). $\qquad\qquad\qquad\qquad\qquad\qquad\square$

4.2 Subsets of a given size

We denote by $\varkappa_m(l)$ the number of subsets of size l in a random partition of an m-element set. If $T_m(k,l)$ is the number of partitions containing k subsets of size l and

$$(2.1) \qquad D_m(x;\, l) := \sum_{k=0}^{[m/l]} T_m(l,\, k)\, x^k$$

then, by formula (0.19) of Chapter 3,

$$(2.2) \qquad \sum_{m=0}^{\infty} D_m(x;\, l)\, \frac{t^m}{m!} \;=\; \exp\left\{ (x-1)\, \frac{t^l}{l!} + e^t - 1 \right\}.$$

Hence we obtain the explicit expression for the generating function of $\varkappa_m(l)$:

$$(2.3) \qquad P_m(x;\, l) \;=\; \frac{D_m(x;\, l)}{D_m(1;\, l)} \;=\; \sum_{k=0}^{[m/l]} \frac{(m)_{kl}\, T_{m-kl}}{k!\, (l!)^k\, T_m}\, (x-1)^k.$$

With the aid of (2.3) it is not difficult to find the binomial moments of $\varkappa_m(l)$:

$$(2.4) \qquad B_k(m;\, l) \;=\; \frac{(m)_{kl}\, T_{m-kl}}{k!\, (l!)^k\, T_m}\,, \qquad\qquad k = 0, 1, \ldots, [m/l].$$

In particular, the mean and variance of $\varkappa_m(l)$ are of the form

$$(2.5) \qquad \begin{cases} \mathbf{E}\,\varkappa_m(l) \;=\; \dfrac{(m)_l\, T_{m-l}}{l!\, T_m}\,, \\[2ex] \mathrm{Var}\,\varkappa_m(l) \;=\; \dfrac{(m)_{2l}\, T_{m-2l}}{(l!)^2\, T_m} + \dfrac{(m)_l\, T_{m-l}}{l!\, T_m} - \left(\dfrac{(m)_l\, T_{m-l}}{l!\, T_m} \right)^2. \end{cases}$$

Recalling formula (0.25) of Chapter 3 we see that, for any fixed k and l, the following asymptotic representation is valid as $m \to \infty$:

$$T_{m-kl} = \frac{1}{\sqrt{r+1}}\, \exp\left\{ (m - kl) \right.$$

$$\left. \times \left(r + \frac{1}{r} - 1 - \frac{klr}{m\,(r+1)} + O\!\left(\frac{1}{m\,r^2}\right) \right) - 1 \right\} (1 + o(1)).$$

By applying this formula, as $m \to \infty$, to (2.4) and (2.5) we deduce, for any fixed l and k, that

$$(2.6) \qquad B_k(m;\, l) \;=\; \frac{1}{k!}\left(\frac{r^l}{l!}\right)^k (1 + o(1)),$$

and

$$
(2.7) \qquad
\begin{cases}
\mathbf{E}\,\varkappa_m(l) = \dfrac{r^l}{l!}\left(1 + o(1)\right), \\[3mm]
\operatorname{Var}\varkappa_m(l) = \dfrac{r^l}{l!}\left(1 + o(1)\right).
\end{cases}
$$

Theorem 2.1 *Let r be the unique real solution of the equation $r\,e^r = m$.*
For any fixed l the distribution of the random variable

$$
\varkappa_m'(l) = \frac{\varkappa_m(l) - r/l!}{\sqrt{r^l/l!}}
$$

converges to the standard normal distribution as $m \to \infty$.

First we prove an auxiliary lemma.

Lemma 1 *As $m \to \infty$, the following asymptotic formula holds:*

$$
(2.8) \quad D_m(x; l) = \left(\frac{m}{R}\right)^m \exp\Bigg\{-m + \frac{m}{R} - 1
$$
$$
- \frac{(x-1)\,R^{l-1}}{(l-1)!} + \frac{(x-1)\,R^l}{l!}\Bigg\}
$$
$$
\times \left[(R+1)\left(1 - \frac{(x-1)\,R^l}{m\,(l-1)!}\right) + \frac{(x-1)\,l\,R^l}{m\,(l-1)!}\right]^{-1/2}
$$
$$
\times \left(1 + o(1)\right),
$$

where $R = R(x)$ is the maximal positive solution of the equation

$$
(2.9) \qquad R\,e^R = m - \frac{(x-1)\,R^l}{(l-1)!}
$$

and $o(1) \to 0$ uniformly for x from a neighborhood $W = \{1 - \delta \le x \le 1 + \delta\}$, $\delta \in (0, 1)$, of the point $x = 1$.

Proof Using Cauchy's formula one can derive from (2.2) that

$$
D_m(x; l) = \frac{m!}{2\pi i} \oint_C \exp\left\{(x-1)\,\frac{z^l}{l!} + e^z - 1\right\} \frac{dz}{z^{m+1}},
$$

where C is an arbitrary closed contour enclosing the origin in the complex plane z. Taking as C the circle of radius R with center at the origin we have

$$
(2.10) \qquad D_m(x; l) = \frac{m!}{2\pi R^m} \exp\left\{e^R - 1 + \frac{(x-1)\,R^l}{l!}\right\} \times J,
$$

where

$$J := \int_{-\pi}^{\pi} \exp\{f(\varphi, R, x)\} \, d\varphi,$$

$$f(\varphi, R, x) := \exp\{Re^{i\varphi}\} - e^R + \frac{(x-1)R^l}{l!} (e^{il\varphi} - 1) - im\varphi.$$

We set $\varepsilon = \exp\{-2R/5\}$ and split up the integral J as follows:

$$J = J_1 + J_2 + J_3 = \int_{-\pi}^{-\varepsilon} + \int_{-\varepsilon}^{\varepsilon} + \int_{\varepsilon}^{\pi}.$$

Now one can find the derivatives of the function $f(\varphi, R, x)$ at $\varphi = 0$:

$$f'(0, R, x) = i\left(Re^R + \frac{(x-1)R^l}{(l-1)!} - m\right),$$

$$f^{(k)}(0, R, x) = i^k\left(\theta e^R + \frac{(x-1)R^l}{l!} l^k\right), \qquad k = 2, 3, \ldots,$$

where

$$\theta^k = R \frac{d}{dR} \theta^{k-1}$$

is the kth iterate of the differential operator $R\frac{d}{dR}$ and θ^0 is the identity operator.

Taking as $R = R(x)$ the maximal positive (for large m) solution of equation (2.9) we expand $f(\varphi, R, x)$ as a Maclaurin series in a neighborhood $-\varepsilon \leq \varphi \leq \varepsilon$, $\varepsilon > 0$, of the point $\varphi = 0$:

$$f(\varphi, R, x) = -\left[(R^2 + R)e^R + \frac{(x-1)l R^l}{(l-1)!}\right] \frac{\varphi^2}{2}$$

$$+ \sum_{k=3}^{\infty} i^k \left[\theta^k e^R + \frac{(x-1)R^l}{l!} l^k\right] \frac{\varphi^k}{k!}.$$

Estimating the last sum of the expression and recalling (2.9), we obtain

$$f(\varphi, R, x) = -\left[(R+1)\left(m - \frac{(x-1)R^l}{(l-1)!}\right) + \frac{(x-1)l R^l}{(l-1)!}\right] \frac{\varphi^2}{2} + o(1),$$

where $o(1) \to 0$, as $m \to \infty$, uniformly for $x \in W$. Now a standard procedure gives

$$J_2 = \sqrt{2\pi} \left[(R+1)\left(m - \frac{(x-1)R^l}{(l-1)!}\right) + \frac{(x-1)l R^l}{(l-1)!}\right]^{-1/2} (1 + o(1)).$$

One can show that J_1 and J_3 decrease exponentially and uniformly in m as $m \to \infty$. And thus representation (2.10) leads to (2.8). $\square$

Proof of Theorem 2.1 From Lemma 1 it follows that the generating function of $\varkappa_m(l)$ admits the representation

$$(2.11) \qquad P_m(x;\, l) = \left(\frac{r}{R}\right)^m \exp\left\{ m\left(\frac{1}{R} - \frac{1}{r}\right)\right.$$

$$\left. - \frac{(x-1)\,R^{l-1}}{(l-1)!} + \frac{(x-1)\,R^l}{l!}\right\}$$

$$\times \left[\frac{R+1}{r+1}\left(1 - \frac{(x-1)\,R^l}{m\,(l-1)!}\right)\right.$$

$$\left. + \frac{(x-1)\,l\,R^l}{(r+1)\,m\,(l-1)!}\right]^{-1/2}\left(1 + o(1)\right),$$

where $r = R(1)$ and $o(1) \to 0$, as $m \to \infty$, uniformly for $x \in W$.

From (2.9) we have

$$R = r\left[1 - \frac{(x-1)\,r^l}{m\,(r+1)\,(l-1)!}\right] + O\!\left(\left(\frac{r^l}{m}\right)^2\right).$$

This equality allows us to simplify the asymptotic representation (2.11) considerably and to establish the formula

$$(2.12) \qquad P_m(x;\, l) = \exp\left\{\frac{(x-1)\,r^l}{l!}\right\}\left(1 + o(1)\right),$$

where $o(1) \to 0$, as $m \to \infty$, uniformly for $x \in W$. From (2.12) we see that, for any fixed x,

$$P_m\left(\exp\left\{\frac{x}{\sqrt{r^l/l!}}\right\};\, l\right) = \exp\left\{x\,\sqrt{r^l/l!} + \tfrac{1}{2}x^2\right\}\left(1 + o(1)\right).$$

Hence we deduce that the moment generating function of the random variable $\varkappa'_m(l)$ converges, as $m \to \infty$, to the moment generating function $e^{x^2/2}$ of the standard normal distribution. Now Theorem 2.1 follows from Curtiss' theorem. $\qquad\qquad\Box$

4.3 Random partitions with constraints on the sizes of blocks

In this section we consider such partitions of an m-element set whose subsets (blocks) belong to a finite set **A**, where

$$(3.1) \qquad \mathbf{A} = \{j_1, j_2, \ldots, j_d\}, \quad 1 \le j_1 < j_2 < \cdots < j_d < \infty, \quad d > 1,$$

and

$$\gcd(j_1, j_2, \ldots, j_d) = 1,$$

that is, the possible sizes of blocks are relatively prime numbers. Let us denote by T^A_{mn} the number of such partitions with n blocks and let T^A_m

be the total number of such partitions. By formula (0.22) of Chapter 3 we have

$$(3.2) \qquad \sum_{m=0}^{\infty} Q_m(x;\ \mathbf{A})\, \frac{t^m}{m!} = \exp\{x\, A(t)\},$$

where

$$(3.3) \qquad Q_m(x;\ \mathbf{A}) := \sum_{n=0}^{m} T_{mn}^A\, x^n,$$

$$(3.4) \qquad A(t) := \sum_{k=1}^{d} \frac{t^{j_k}}{j_k!}.$$

Clearly,

$$(3.5) \qquad T_m^A = Q_m(1;\ \mathbf{A}).$$

Applying the corollary to Theorem 4.1 of Chapter 1 we obtain, as $m \to \infty$, the asymptotic representation

$$(3.6) \qquad T_m^A = \frac{m!\, e^{A(r)}}{r^m\, \sqrt{2\,\pi\, A_2(r)}} \left(1 + o(1)\right),$$

where, for large m, r is the unique positive root of the equation

$$(3.7) \qquad \sum_{k=1}^{d} \frac{r^{j_k}}{(j_k - 1)!} = m$$

and

$$A_2(r) := \sum_{k=1}^{d} \frac{j_k\, r^{j_k}}{(j_k - 1)!}.$$

We introduce the notations

$$u(t) := \sum_{k=1}^{d-1} \frac{t^{j_k}}{j_k!}, \qquad\qquad \chi := t\, \frac{d}{dt},$$

$$u_j(t) := \chi^j\, u(t), \qquad\qquad A_j(t) := \chi^j\, A(t), \qquad j = 1, 2, \ldots.$$

In addition, put

$$M_m := u(r) - \frac{A_1(r)}{A_2(r)}\, u_1(r) + \frac{1}{j_d}\left(u_1(r) - \frac{A_1(r)}{A_2(r)}\, u_2(r)\right) \frac{u_1(r)}{m - u_1(r)},$$

$$\sigma_m^2 := u(r) - \frac{2}{j_d}\, u_1(r) + \frac{1}{j_d^2}\, u_2(r),$$

where r is the unique positive root of equation (3.7).

Theorem 3.1 *Let $\xi_m(\mathbf{A})$ be the number of blocks in a random equiprobable partition whose blocks satisfy conditions (3.1). Then the distribution of the random variable*

$$\xi'_m(\mathbf{A}) = \frac{\xi_m(\mathbf{A}) - m\, j_d^{-1} - M_m}{\sigma_m}$$

converges to the standard normal distribution as $m \to \infty$.

Proof It follows from (3.2)–(3.5) that the generating function of $\xi_m(\mathbf{A})$ is of the form

$$f_m(x) = Q_m(x; \mathbf{A}) / Q_m(1; \mathbf{A}).$$

Conditions (3.1) allow us to use Theorem 4.3 of Chapter 1 whose reformulation to the case of interest completes the proof of Theorem 3.1. □

The mean and variance of $\xi_m(\mathbf{A})$ admit the following asymptotic representations as $m \to \infty$:

$$\mathbf{E}\,\xi_m(\mathbf{A}) = \frac{m}{j_d} + \sum_{k=1}^{d-1} \left(1 - \frac{k}{j_d}\right) \cdot \frac{r^k}{j_k!}(1 + o(1)),$$

$$\mathrm{Var}\,\xi_m(\mathbf{A}) = \sum_{k=1}^{d-1} \left(1 - \frac{k}{j_d}\right)^2 \cdot \frac{r^k}{j_k!}(1 + o(1)).$$

If d is held fixed, the blocks of maximal size j_d give the main contribution to the mean as $m \to \infty$.

4.4 Multivariate limit theorem

In Section 2 of this chapter we established the asymptotic normality of the random variable

$$\frac{\varkappa_m(l) - \lambda_l}{\sqrt{\lambda_l}}$$

as $m \to \infty$, where $\varkappa_m(l)$ is the number of blocks of size l in a random partition of an m-element set, $\lambda_l = r^l/l!$ and $r\,e^r = m$. To simplify the notation we write $\varkappa_l = \varkappa_m(l)$. Let us consider a k-dimensional random variable

$$\varkappa^{(k)} = (\varkappa_{i_1}, \varkappa_{i_2}, \dots, \varkappa_{i_k}), \qquad 1 \le i_1 < i_2 < \cdots < i_k \le m.$$

The main goal of the arguments to follow is to establish the local asymptotic normality with unit variances and independent components of the random vector

$$(4.1) \qquad \left((\varkappa_{i_1} - \lambda_{i_1})/\sqrt{\lambda_{i_1}}, \, \dots, \, (\varkappa_{i_k} - \lambda_{i_k})/\sqrt{\lambda_{i_k}}\right), \qquad m \to \infty.$$

First we prove a number of auxiliary lemmas.

Lemma 1 *For* $1 \leq i_1 < i_2 < \cdots < i_k \leq m$, $\nu = i_1 j_1 + \cdots + i_k j_k$, $m = 1, 2, \ldots$, *the formulae*

(4.2)
$$\mathbf{P}\left\{\varkappa_{i_1} = j_1, \ldots, \varkappa_{i_k} = j_k\right\}$$

$$= \frac{(m)_\nu}{(i_1!)^{j_1} \cdots (i_k!)^{j_k} \, j_1! \cdots j_k!} \cdot \frac{D_{m-\nu}}{T_m}$$

hold, where T_m *are Bell's numbers and the numbers* D_m *are determined by the generating function*

(4.3)
$$\sum_{m=0}^{\infty} D_m \frac{t^m}{m!} = \exp\left\{ e^t - 1 - \sum_{l=1}^{k} \frac{t^{i_l}}{i_l!} \right\}.$$

Proof The number of ways of choosing ν elements from m given elements and constructing from them partitions of the secondary specification $\left[i_1^{j_1}, i_2^{j_2}, \ldots, i_k^{j_k} \right]$ is equal to $(m)_\nu \, [(i_1!)^{j_1} \cdots (i_k!)^{j_k} \, j_1! \cdots j_k!]^{-1}$. To find the required probability it suffices to multiply the quantity obtained by $D_{m-\nu}$, the number of partitions of an $(m-\nu)$-set with no subsets of sizes $i_1, i_2, \ldots, i_k$, and to divide the result by T_m. Hence formula (4.2) follows. Relation (4.3) for the generating function of the sequence $\{D_m\}$ is a particular case of formula (0.21) from Chapter 3, with $\mathbf{A} = \mathbb{N} \setminus \{i_1, i_2, \ldots, i_k\}$, where $\mathbb{N}$ is, as usual, the set of all positive integers. $\square$

Lemma 2 *Let* $\varphi(t)$ *be a polynomial of fixed degree in* t *and*

$$\sum_{m=0}^{\infty} D_m^\varphi \frac{t^m}{m!} = \exp\left\{ e^t - 1 - \varphi(t) \right\}.$$

Then

(4.4)
$$D_m^\varphi = \frac{m! \, \exp\left\{ e^R - 1 - \varphi(R) \right\}}{R^m \left(2\pi R \left(R + 1 \right) e^R \right)^{1/2}} \left(1 + o(1) \right), \qquad m \to \infty,$$

where, for large m, R *is the maximal real root of the equation*

(4.5)
$$e^R - \varphi'(R) = \frac{m}{R}.$$

Here the prime denotes differentiation.

Proof By Cauchy's formula,

$$D_m^\varphi = \frac{m!}{2\pi i} \oint_C \exp\left\{ e^z - 1 - \varphi(z) \right\} \frac{dz}{z^{m+1}}.$$

Taking as the contour C the circle of radius R with center at the origin, where R is the solution of equation (4.5), we obtain

$$D_m^\varphi = \frac{m!}{2\pi R^m} \exp\left\{ e^R - 1 - \varphi(R) \right\} \times J ,$$

where

$$J := \int_{-\pi}^{\pi} \exp\left\{ f(R,\varphi) \right\} d\varphi ,$$

$$f(R,\varphi) := \left[\exp\{R e^{i\theta}\} - e^R \right] - \left[\varphi(R e^{i\theta}) - \varphi(R) \right] - i m \theta .$$

We set $\varepsilon = \exp\{-2R/5\}$ and split up the integral J into three integrals: $J = J_1 + J_2 + J_3$ with integration limits $[-\pi, -\varepsilon]$, $[-\varepsilon, \varepsilon]$ and $[\varepsilon, \pi]$ respectively. One can show that

$$J_2 = \frac{\sqrt{2\pi}}{\sqrt{R (R+1) e^R}} \left(1 + o(1) \right) ,$$

$$|J_l| = o\left(bigl(R (R+1) e^R)^{-1/2} \right) , \qquad l = 1, 3 .$$

Combining the estimates obtained we prove Lemma 2. $\square$

Lemma 3 $m \to \infty$, *the asymptotic representation*

$$(4.6) \qquad D_{m-\nu}^\varphi = \frac{r^\nu\, e^{-\varphi(r)}}{(m)_\nu}\, T_m \left(1 + o(1) \right)$$

holds for $\nu = O\left(r\, \varphi'(r) \right)$ *and* $r\, e^r = m$.

Proof   From formula (4.4) it follows that

$$D_{m-\nu}^\varphi = \frac{(m-\nu)!\, \exp\left\{ e^{R_\nu} - 1 - \varphi(R_\nu) \right\}}{R_\nu^{m-\nu} (2\pi R_\nu (R_\nu + 1) e^{R_\nu})^{1/2}} \left(1 + o(1) \right) ,$$

where, for large m, R_ν is the maximal root of the equation

$$(4.7) \qquad e^{R_\nu} - \varphi'(R_\nu) = \frac{m-\nu}{R_\nu} .$$

Thus, the following relation is valid:

$$(4.8) \qquad \frac{D_{m-\nu}^\varphi}{T_m} = \frac{r^m}{(m)_\nu R_\nu^{m-\nu}} \left(\frac{r (r+1) e^r}{R_\nu (R_\nu + 1)\, e^{R_\nu}} \right)^{1/2}$$

$$\times \exp\left\{ e^{R_\nu} - e^r - \varphi(R_\nu) \right\} \left(1 + o(1) \right) .$$

By the equality $r\,e^r = m$ we deduce from (4.7) that

$$R_\nu = r + \frac{r^2\,\varphi'(r)}{(r+1)\,(m-\nu)} - \frac{r\,\nu}{(r+1)\,m} + O\!\left(\left(\frac{r\,\varphi'(r)}{m}\right)^2\right).$$

Therefore,

$$R_\nu^{m-\nu} = r^{m-\nu}\,\exp\left\{\frac{r\,\varphi'(r)}{r+1} - \frac{\nu}{r+1} + O\!\left(\frac{r\,\varphi'^2(r)}{m}\right)\right\}.$$

For the inverse quantity R_ν we obtain the representation

$$\frac{1}{R_\nu} = \frac{1}{r} - \frac{\varphi'(r)}{(r+1)\,(m-\nu)} + \frac{\nu}{r\,(r+1)\,m} + O\!\left(r\left(\frac{\varphi'(r)}{m}\right)^2\right),$$

yielding

$$(4.9)\qquad e^{R_\nu} - e^r = \frac{r}{r+1}\,\varphi'(r) - \frac{\nu}{r+1} + O\!\left(\frac{r\,\varphi'^2(r)}{m}\right).$$

Combining (4.8) and (4.9) with the asymptotic expansion

$$\varphi\,(R_\nu) = \varphi(r) + O\!\left(\frac{r\,\varphi'(r)}{m}\right),$$

we conclude that Lemma 3 is valid.   □

We now formulate and prove a local limit theorem.

Theorem 4.1 $m \to \infty$, the asymptotic relation

$$(4.10)\quad \mathbf{P}\left\{\frac{\varkappa_{i_1} - \lambda_{i_1}}{\sqrt{\lambda_{i_1}}} = \mu_1,\ \dots\ ,\ \frac{\varkappa_{i_k} - \lambda_{i_k}}{\sqrt{\lambda_{i_k}}} = \mu_k\right\}$$

$$= \frac{1}{(2\pi)^{k/2}\,(\lambda_{i_1}\,\lambda_{i_2}\,\cdots\,\lambda_{i_k})^{1/2}}\,\exp\left\{-\tfrac{1}{2}\sum_{s=1}^{k}\mu_s^2\right\}(1+o(1))$$

holds uniformly over $\{|\,\mu_i\,| \le C < \infty,\ i = 1, 2, \dots, k\}$, where $\lambda_l = r^l/l!$, r is the unique positive root of the equation $r\,e^r = m$ and C is a positive constant.

Proof   We introduce the new variables μ_s, $0 \le \mu_s \le C < \infty$, by the equalities

$$j_s = \lambda_{i_s} + \mu_s\,\sqrt{\lambda_{i_s}}\,,\qquad\qquad s = 1, 2, \dots, k\,.$$

Put $H := (i_1!)^{j_1}\cdots(i_k!)^{j_k}\,j_1!\cdots j_k!$. Since $r \to \infty$ as $m \to \infty$, the estimate

$$\log H = \tfrac{1}{2}k\log 2\pi + \log r \sum_{s=1}^{k} i_s\,\lambda_{i_s} - \sum_{s=1}^{k}\lambda_{i_s}$$

$$+ \tfrac{1}{2}\sum_{s=1}^{k}\log\lambda_{i_s} + \log r \sum_{s=1}^{k} i_s\,\mu_s\,\lambda_{i_s}^{1/2} + \tfrac{1}{2}\sum_{s=1}^{k}\mu_s^2 + o(1)$$

is valid, where, as $m \to \infty$, $o(1)$ tends to 0 uniformly over $\{|\mu_i| \leq C,\, i = 1, 2, \ldots, k\}$. Writing $\varphi(t)$ in the form

$$\varphi(t) = \sum_{l=1}^{k} \frac{t^{i_l}}{i_l!},$$

we find that

$$\nu = r\,\varphi'(r) + \sum_{s=1}^{k} i_s \,\mu_s \,\lambda_{i_s}^{1/2}, \qquad \varphi(r) = \sum_{s=1}^{k} \lambda_{i_s},$$

allowing us to establish the relation

$$(4.11) \quad H = (2\pi)^{k/2} \,(\lambda_{i_1} \cdots \lambda_{i_k})^{1/2} \, r^{\nu} \, \exp\left\{-\varphi(r) + \tfrac{1}{2} \sum_{s=1}^{k} \mu_s^2\right\} (1 + o(1)).$$

According to Lemma 1,

$$\mathbf{P}\left\{ \frac{\varkappa_{i_1} - \lambda_{i_1}}{\sqrt{\lambda_{i_1}}} = \mu_1, \ \ldots, \ \frac{\varkappa_{i_k} - \lambda_{i_k}}{\sqrt{\lambda_{i_k}}} = \mu_k \right\} = \frac{(m)_{\nu}}{H} \cdot \frac{D_{m-\nu}}{T_m}.$$

Finally, combining (4.2), (4.6) and (4.11) we obtain (4.10).   □

From the local theorem one can deduce the following integral theorem by a standard procedure.

Theorem 4.2   *For any squarable k-dimensional domain* $\mathbf{L}_k$

$$\lim_{m \to \infty} \mathbf{P}\left\{ \left(\frac{\varkappa_{i_1} - \lambda_{i_1}}{\sqrt{\lambda_{i_1}}}, \ldots, \frac{\varkappa_{i_k} - \lambda_{i_k}}{\sqrt{\lambda_{i_k}}} \right) \in \mathbf{L}_k \right\}$$

$$= \frac{1}{(2\pi)^{k/2}} \int \cdots \int_{\mathbf{L}_k} \exp\left\{ -\tfrac{1}{2} \sum_{s=1}^{k} y_s^2 \right\} dy_1\, dy_2 \, \ldots \, dy_k.$$

4.5   Maximum and minimum subsets of a partition

We denote by $N_k^{(m)}$ and $M_k^{(m)}$ the numbers of partitions of an m-element set, all of whose blocks have sizes exceeding k and not exceeding k respectively. From formulae (0.14) and (0.21) of Chapter 3 we know that

$$(5.1) \quad \begin{aligned} \sum_{m=0}^{\infty} N_k^{(m)} \frac{t^m}{m!} &= \exp\left\{ e^t - 1 - \varphi_k(t) \right\}, & N_0^{(m)} &= T_m, \\ \sum_{m=0}^{\infty} M_k^{(m)} \frac{t^m}{m!} &= \exp\left\{ \varphi_k(t) \right\}, & M_0^{(0)} &= 1, \end{aligned}$$

where

$$(5.2) \qquad \varphi_0(t) = 0, \qquad \varphi_k(t) = \sum_{j=1}^{k} \frac{t^j}{j!}, \qquad k = 1, 2, \ldots,$$

and T_m are Bell's numbers.

We denote by ν_m and μ_m the sizes of the minimum and maximum blocks, respectively, in a random equiprobable partition of an m-element set. It is evident that

$$\mathbf{P}\{\nu_m > k\} = N_k^{(m)}/T_m, \qquad \mathbf{P}\{\mu_m \le k\} = M_k^{(m)}/T_m.$$

First we investigate the asymptotic distribution of ν_m as $m \to \infty$. To this end we prove a more general statement.

Theorem 5.1 *For any fixed l,*

$$\lim_{m \to \infty} \mathbf{P}\{\varkappa_m(l) > 0\} = 1.$$

Proof For any nonnegative random variable $\varkappa$ the inequality of Chebyshev's type holds:

$$\mathbf{P}\{\varkappa = 0\} \le \operatorname{Var} \varkappa \cdot (\mathbf{E}\varkappa)^{-2}.$$

Applying this inequality we obtain

$$\mathbf{P}\{\varkappa_m(l) > 0\} \ge 1 - \operatorname{Var}\varkappa_m(l) \cdot (\mathbf{E}\varkappa_m(l))^{-2}.$$

Recalling (2.7) we conclude that, for large m, the estimate

$$\mathbf{P}\{\varkappa_m(l) > 0\} \ge 1 - \frac{l!}{r^l}, \qquad r\,e^r = m,$$

is valid. Hence the theorem follows. $\square$

Roughly speaking, Theorem 5.1 states that, for any fixed l, any random partition has blocks of size l with probability close to 1 as $m \to \infty$. In particular, it has single element blocks. Consequently,

$$\lim_{m \to \infty} \mathbf{P}\{\nu_m = 1\} = 1,$$

that is, the limiting distribution of ν_m is degenerate.

Now we proceed to study the asymptotic distribution of the random variable μ_m as $m \to \infty$. We introduce the notation

$$\widetilde{\Delta}_k := \sum_{l=k}^{\infty} \frac{\widetilde{R}^j}{j!},$$

where, for fixed m and k, $\tilde{R} = \tilde{R}(m, k)$ is the maximum root of the following equation:

$$(5.3) \qquad \tilde{R}\, \varphi'_k (\tilde{R}) = m\,.$$

Lemma 1 *As $m \to \infty$ and $k = o(r)$, $r\, e^r = m$, we have*

$$(5.4) \qquad \mathbf{P}\, \{\mu_m \leq k\} = \exp \left\{ -\tilde{\Delta}_k + \frac{\tilde{R}^k}{k!} + O\left(\frac{\tilde{R}\, \tilde{\Delta}_k^2}{m} \right) \right\}.$$

Proof We note that the asymptotic representation is useful for only sufficiently large k. Applying the method used in the proof of formula (4.4) and taking into account formulae (5.1) and (5.2) for $k = o(r)$, we find that

$$(5.5) \quad \mathbf{P}\, \{\mu_m \leq k\} = \exp \left\{ m \left(\frac{1}{\tilde{R}} - \frac{1}{r} + \log \frac{r}{\tilde{R}} \right) + \frac{\tilde{R}^k}{k!} \right\} (1 + o(1))\,.$$

From equation (5.3) one can deduce the expression

$$\tilde{R} = r + \frac{r^2\, \tilde{\Delta}_k}{(r + 1)\, m} + O\left(\left(\frac{\tilde{R}\, \tilde{\Delta}_k}{m} \right)^2 \right),$$

yielding

$$m \left(\frac{1}{\tilde{R}} - \frac{1}{r} + \log \frac{r}{\tilde{R}} \right) = -\tilde{\Delta}_k + O\left(\frac{\tilde{R}\, \tilde{\Delta}_k^2}{m} \right).$$

Substituting the estimate obtained into (5.5) we arrive at the assertion of the lemma. □

We put

$$R = \tilde{R}(m, k_0)\,,$$

where

$$k_0 := [\, e\, r - \log \sqrt{2\pi r} - \log (e - 1)\,]$$

and square brackets denote the integral part of the corresponding real number. Let

$$\alpha = \log \sqrt{2\pi e r} + \log (e - 1)$$

and let $\Delta(R, x)$ and $\gamma(R, x)$ be the values of $\Delta_k = \sum_{j=k}^{\infty} R^j / j!$ and $R^k / k!$ for $k = [\, e\, R - \alpha\,] + [x]$. We write k in the form $k = e\, R - \alpha - \varepsilon + [x]$, where $\varepsilon = \{e\, R - \alpha\}$ is the fractional part of $e\, R - \alpha$.

Lemma 2 *The asymptotic representations*

$$(5.6) \qquad \Delta(R, x) = \exp \{ 1 + \varepsilon - [x] \} (1 + o(1))\,,$$

$$(5.7) \qquad \gamma(R, x) = (e - 1)\, \exp \{ \varepsilon - [x] \} (1 + o(1))$$

are valid as $m \to \infty$, where $o(1) \to 0$ uniformly over $\{ |x| \leq \log \sqrt{r} \}$.

Proof We represent the expression

$$\Delta(R,x) = \sum_{j=eR-\alpha-\varepsilon+[x]}^{\infty} \frac{R^j}{j!}$$

as a sum $\Delta(R,x) = S_1 + S_2$, where

$$S_1 := \sum_{j=eR-\alpha-\varepsilon+[x]}^{eR-\alpha-\varepsilon+[x]+3\alpha} \frac{R^j}{j!}, \qquad S_2 := \sum_{j=eR-\alpha-\varepsilon+[x]+3\alpha+1}^{\infty} \frac{R^j}{j!}.$$

Obvious estimates lead to

$$S_1 = \frac{1}{\sqrt{2\pi eR}} \sum_{j=0}^{3\alpha} \exp\left\{-(j-\alpha-\varepsilon+[x])\right\}\left(1+o(1)\right),$$

$$S_2 = O\left(\frac{e^{-a}}{\sqrt{2\pi eR}}\right)$$

uniformly over $\{\,|x| \le \log\sqrt{r}\,\}$. Moreover, it is not difficult to show that

$$S_1 = \frac{\exp\left\{\alpha+\varepsilon+1-[x]\right\}}{\sqrt{2\pi eR}\,(e-1)}\left(1+o(1)\right).$$

Since $e^{\alpha}/(\sqrt{2\pi eR}\,(e-1)) = 1+o(1)$ since formula (5.6) follows immediately from the estimates obtained for S_1 and S_2.

Using Stirling's formula we can easily deduce the representation

$$\gamma(R,x) = \frac{\exp\left\{\alpha+\varepsilon-[x]\right\}}{\sqrt{2\pi eR}}\left(1+o(1)\right),$$

from which equality (5.7) follows, thereby completing the proof of the lemma. ◻

We now consider the distribution of the random variable

$$\bar{\mu}_m := \mu_m - \left[\,er - \log\sqrt{2\pi er} - \log\,(e-1)\,\right]$$

and denote by δ the fractional part of the number

$$er - \log\sqrt{2\pi er} - \log\,(e-1).$$

Theorem 5.2 *If $m \to \infty$ then*

$$\mathbf{P}\left\{\bar{\mu}_m < x\right\} - \exp\left\{-e^{-[x]+\delta}\right\} \longrightarrow 0$$

uniformly over any bounded domain.

Proof Lemma 1 with $k = [eR - \log\sqrt{2\pi eR} - \log(e-1)] + [x]$ gives the estimate

$$\mathbf{P}\{\mu_m \leq k\} = \exp\{-\Delta_k + R^k/k!\}(1 + o(1)), \qquad m \to \infty,$$

where $o(1) \to 0$ uniformly over $\{x\colon |x| \leq \log\sqrt{r}\}$. Further, applying Lemma 2, we have

$$(5.8) \quad \mathbf{P}\left\{\mu_m \leq [eR - \log\sqrt{2\pi eR} - \log(e-1)] + [x]\right\}$$

$$= \exp\{-e^{-[x]+\varepsilon}\}(1 + o(1)),$$

where $o(1) \to 0$, as $m \to \infty$, uniformly over any bounded domain. Arguing as in the proof of Lemma 1 we see that the uniform estimate

$$R = r + O\left(\frac{\log m}{m}\right)$$

holds. Since the left-hand side of (5.8) is a step function with jumps at integer points, we obtain

$$\mathbf{P}\{\bar{\mu}_m < x\} = \exp\{-e^{-[x]+\delta}\}(1 + o(1)).$$

Hence the validity of Theorem 5.2 follows. □

The meaning of Theorem 5.2 is that, for large m, the distribution of μ_m is concentrated within a neighborhood of the point

$$[er - \log\sqrt{2\pi er} - \log(e-1)]$$

and in the domain it is closed to the double exponential distribution. The function $\delta = \delta(m)$, $0 \leq \delta < 1$, has no limit as $m \to \infty$. The asymptotic behavior of δ depends on the form of the subsequence through which parameter m runs as $m \to \infty$. The concentration point of the distribution of μ_m is closed to the mean of the random variable as $m \to \infty$. To justify this statement we will study the behavior of $\mathbf{E}\,\mu_m$ as $m \to \infty$. We claim that

$$(5.9) \qquad \mathbf{E}\,\mu_m = er - \log\sqrt{r} + O(1), \qquad m \to \infty.$$

The scheme for proving this formula is as follows. First we use the representation

$$\mathbf{E}\,\mu_m = \sum_{j=1}^{m} \left[1 - \mathbf{P}\{\mu_m \leq j\}\right]$$

and show

$$\mathbf{E}\,\mu_m = \sum_{0 \leq j \leq [eR - \log(e-1)]} \left[1 - \mathbf{P}\{\mu_m \leq j\}\right] + O(1), \qquad m \to \infty.$$

Further, we state that

$$\sum_{0 \le j \le [eR - \log(e-1)]} \mathbf{P}\{\mu_m \le j\}$$

$$= \sum_{[eR - \beta] \le j \le [eR - \log(e-1)]} \mathbf{P}\{\mu_m \le j\} + O(1),$$

where $\beta = \log\sqrt{2\pi eR} + \log(e-1)$. Finally, applying Theorem 5.2 yields

$$\sum_{[eR - \beta] \le j \le [eR - \log(e-1)]} \mathbf{P}\{\mu_m \le j\} = \log\sqrt{R} + O(1).$$

Collecting the estimates above and passing to the limit as $m \to \infty$, we obtain (5.9).

4.6 Partitions with labeled blocks

4.6.1 A bicycle race problem

Consider a bicycle race with m participants who split into groups during the competition, and assume that such partitioning into groups is random (the race is not a team competition), corresponding to a random decomposition of a given m-element set into blocks. Suppose that the maintenance of the cycles is organized centrally as is the number of cars to be used by the repair teams during the race. Servicing the race will be optimized if each group of racers is accompanied, on average, by one car. Provided that the partition of the racers into groups is random and equiprobable, the mean number of cars required is, according to formula (1.4),

$$\mathbf{E}\,\xi_m = \frac{T_{m+1}}{T_m} - 1.$$

For large m this quantity amounts asymptotically to $m/\log m$, which is a considerable number and it is hardly expedient to have such a large number of servicing cars. There is also a more obvious reason to reduce the number of cars: malfunction of bicycles occurs only with a certain probability and, given equal reliability of the bicycles, the probability of the occurrence of a malfunction within a group of bicycles is an increasing function of the number of racers in the group under consideration. But what about the correct answer concerning the number of cars? To find it we have to consider random partitions with labeled blocks.

4.6.2 Random AΛ-partitions with labeled blocks

Let sequences $\mathbf{A}$ and $\mathbf{\Lambda}$ be given such that

$$\mathbf{A} \subseteq \mathbb{N} = \{1, 2, \ldots\}, \quad \mathbf{\Lambda} = \{\Lambda_1, \Lambda_2, \ldots\}, \quad \Lambda_j \subseteq N_0 = \{0, 1, \ldots\},$$
$$j = 1, 2, \ldots.$$

A partition of an m-element set $\mathbf{X}$ is called an $\mathbf{A}\Lambda$-*partition* if the sizes of blocks in the partition belong to the sequence $\mathbf{A}$ and the number β_j of blocks of size j is an element of the sequence Λ_j, $j = 1, 2, \ldots$. By definition we assume that $\Lambda_j = \{0\}$ if $j \notin \mathbf{A}$. The number of $\mathbf{A}\Lambda$-partitions having the secondary specification

$$\left[[1^{\beta_1} 2^{\beta_2} \ldots m^{\beta_m}]\right], \qquad 1\beta_1 + 2\beta_2 + \cdots + m\beta_m = m,$$

is

$$T_m(\beta_1, \ldots, \beta_m; \mathbf{A}, \Lambda) = \begin{cases} m! \quad [(1!)^{\beta_1} \ldots (m!)^{\beta_m} \beta_1! \ldots \beta_m!]^{-1} \\ \qquad \text{if } \beta_j \in \Lambda_j \text{ for all } j = 1, 2, \ldots, m; \\ 0 \quad \text{otherwise}. \end{cases}$$

The *indicator* of $\mathbf{A}\Lambda$-partitions is defined by the relation

(6.1) $\quad C(x_1, x_2, \ldots, x_m; \mathbf{A}, \Lambda)$

$$= \sum_{\sigma_j \, j\beta_j = m} T_m(\beta_1, \ldots, \beta_m; \mathbf{A}, \Lambda) \, x_1^{\beta_1} x_2^{\beta_2} \ldots x_m^{\beta_m}.$$

Using (6.1) one can obtain the generating function with an infinite number of variables, called the *numerator* of $\mathbf{A}\Lambda$-partitions:

(6.2) $\quad F(t; x_1, x_2, \ldots, \mathbf{A}, \Lambda) = \sum_{m=0}^{\infty} C(x_1, x_2, \ldots, x_m; \mathbf{A}, \Lambda) \, \dfrac{t^m}{m!}$

$$= \prod_{j \in A} \sum_{\beta_j \in \Lambda_j} \left(\frac{x_j t^j}{j!} \right)^{\beta_j} \frac{1}{\beta_j!}.$$

Let $T_m(\mathbf{A}, \Lambda)$ be the number of $\mathbf{A}\Lambda$-partitions of an m-element set and $T_{mn}(\mathbf{A}, \Lambda)$ be the number of such partitions having exactly n blocks. Then

(6.3) $\quad \displaystyle\sum_{m=0}^{\infty} T_m(\mathbf{A}, \Lambda) \frac{t^m}{m!} = \prod_{j \in A} \sum_{\beta_j \in \Lambda_j} \left(\frac{t^j}{j!} \right)^{\beta_j} \frac{1}{\beta_j!},$

(6.4) $\quad \displaystyle\sum_{m=0}^{\infty} \sum_{n=0}^{m} T_{mn}(\mathbf{A}, \Lambda) \frac{t^m}{m!} x^n = \prod_{j \in A} \sum_{\beta_j \in \Lambda_j} \left(\frac{xt^j}{j!} \right)^{\beta_j} \frac{1}{\beta_j!}.$

We fix $\mathbf{A}$ and Λ and consider the uniform probability distribution on the set of $\mathbf{A}\Lambda$-partitions. Let a partition appear as a result of random choice according to the uniform distribution. We define a random procedure for labeling the blocks of the partition. Each block of size k, $k = 1, 2, \ldots,$

of the $\mathbf{A}\boldsymbol{\Lambda}$-partition is labeled with probability p_k and remains unlabeled with probability $q_k = 1 - p_k$ independently of other blocks.

We study the random vector $\boldsymbol{\xi}_m(\mathbf{A}, \boldsymbol{\Lambda}) = (\xi^{(1)}, \xi^{(2)}, \dots, \xi^{(m)})$, where $\xi^{(k)}$ is the number of labeled blocks of size k (k-subsets) in a random $\mathbf{A}\boldsymbol{\Lambda}$-partition of an m-element set $\mathbf{X}$. If $\mathbf{k} = (k_1, k_2, \dots, k_m)$ then

$$\mathbf{P}\{\boldsymbol{\xi}_m(\mathbf{A}, \boldsymbol{\Lambda}) = \mathbf{k}\} = \frac{1}{T_m(\mathbf{A}, \boldsymbol{\Lambda})} \sum_{\Sigma_j \, j \, \beta_j = m} T_m(\beta_1, \dots, \beta_m; \mathbf{A}, \boldsymbol{\Lambda})$$

$$\times \prod_{i=1}^{m} \binom{\beta_i}{k_i} p^{k_i} q^{\beta_i - k_i}.$$

The generating function

$$f(x_1, x_2, \dots, x_m; \mathbf{A}, \boldsymbol{\Lambda})$$

$$= \sum_{k_1=0}^{m} \cdots \sum_{k_m=0}^{m} \mathbf{P}\{\boldsymbol{\xi}_m(\mathbf{A}, \boldsymbol{\Lambda}) = \mathbf{k}\} x_1^{k_1} x_2^{k_2} \cdots x_m^{k_m}$$

has the following representation in terms of $\mathbf{A}\boldsymbol{\Lambda}$-partitions:

$$f(x_1, x_2, \dots, x_m; \mathbf{A}, \boldsymbol{\Lambda})$$

$$= \frac{1}{T_m(\mathbf{A}, \boldsymbol{\Lambda})} C(p_1 x_1 + q_1, p_2 x_2 + q_2, \dots, p_m x_m + q_m; \mathbf{A}, \boldsymbol{\Lambda}).$$

Using expression (6.2) for the generating function of the indicator of $\mathbf{A}\boldsymbol{\Lambda}$-partitions we obtain

(6.5) $$\sum_{m=0}^{\infty} T_m(\mathbf{A}, \boldsymbol{\Lambda}) f(x_1, x_2, \dots, x_m; \mathbf{A}, \boldsymbol{\Lambda}) \frac{t^m}{m!}$$

$$= \prod_{j \in A} \sum_{\beta_j \in \Lambda_j} \left((p_j x_j + q_j) \frac{t^j}{j!} \right)^{\beta_j} \frac{1}{\beta_j!}.$$

Let $\xi(m)$ be the number of labeled blocks in a random $\mathbf{A}\boldsymbol{\Lambda}$-partition, let η_m be the total number of elements of $\mathbf{X}$ belonging to the labeled blocks, and let $f_m(x; \mathbf{A}, \boldsymbol{\Lambda})$ and $\varphi_m(x; \mathbf{A}, \boldsymbol{\Lambda})$ be the generating functions of the random variables $\xi(m)$ and η_m respectively. We note that

$$\xi(m) = \xi^{(1)} + \xi^{(2)} + \cdots + \xi^{(m)},$$

$$\eta_m = 1\,\xi^{(1)} + 2\,\xi^{(2)} + \cdots + m\,\xi^{(m)}.$$

Hence it follows that

$$(6.6) \qquad f_m(x; \mathbf{A}, \Lambda) = f(x, x, \dots, x; \mathbf{A}, \Lambda),$$

$$(6.7) \qquad \varphi_m(x; \mathbf{A}, \Lambda) = f(x, x^2, \dots, x^m; \mathbf{A}, \Lambda).$$

Moreover, if $1 \le i_1 < i_2 < \cdots < i_k \le m$ and $\left(\xi^{(i_1)}, \xi^{(i_2)}, \dots, \xi^{(i_k)}\right)$ is the corresponding random vector then its generating function is as follows:

$$f_m(x_{i_1}, x_{i_2}, \dots, x_{i_k}; \mathbf{A}, \Lambda) = f(1, \dots, 1, x_{i_1}, 1, \dots, 1, x_{i_k}, 1, \dots; \mathbf{A}, \Lambda),$$

where the units occur only at positions other than $i_1, i_2, \dots, i_k$. In particular, the generating function of $\xi^{(l)}$ is

$$(6.8) \qquad f_m^{(l)}(x; \mathbf{A}, \Lambda) = f(1, \dots, x, \dots, 1; \mathbf{A}, \Lambda),$$

where x is the lth term.

Let us consider some particular examples of specifying $\mathbf{A}$ and Λ.
1. Let $\mathbf{A} = \mathbb{N} = \{1, 2, \dots\}$ and $\Lambda_j = \mathbb{N}_0 = \{0, 1, 2, \dots\}$, $j = 1, 2, \dots$. The corresponding sequence of generating functions $f(x_1, x_2, \dots, x_m)$, $m = 0, 1, 2, \dots$, satisfies the relationship

$$(6.9) \qquad \sum_{m=0}^{\infty} T_m f(x_1, x_2, \dots, x_m) \frac{t^m}{m!} = \exp\left\{ e^t - 1 + \sum_{k=1}^{\infty} p_k (x_k - 1) \frac{t^k}{k!} \right\},$$

where T_m, $m = 0, 1, \dots$, are Bell's numbers.

Remark It is useful to note that

$$T_m f(x_1, x_2, \dots, x_m) = \operatorname{coeff}_{t^m/m!} \exp\left\{ \sum_{k=1}^{m} (p_k x_k + q_k) \frac{t^k}{k!} \right\}.$$

This equality allows us to assume that, generally speaking, $p_k = p_k(m)$, $k = 1, 2, \dots, m$, and thus to consider the scheme for series of labeling procedures.

If $f_m(x)$ is the generating function of $\xi(m)$ and $g_m(x)$ is the generating function of its binomial moments then, in view of the equality $g_m(x) = f_m(x+1)$, we have

$$(6.10) \qquad \sum_{m=0}^{\infty} T_m g_m(x) \frac{t^m}{m!} = \exp\left\{ e^t - 1 + x \sum_{k=1}^{\infty} p_k \frac{t^k}{k!} \right\}.$$

Hence one can derive the binomial moments:

$$(6.11) \qquad B_k(m) = \frac{1}{k! \, T_m} \sum_{l=k}^{m} \binom{m}{l} T_{m-l}$$

$$\times \sum_{\substack{j_1 + \cdots + j_k = l \\ j_l > 0}} \frac{l!}{j_1! \cdots j_k!} \, p_{j_1} \cdots p_{j_k}.$$

Making use of the inversion formula we derive the exact distribution of $\xi(m)$:

$$\mathbf{P}\{\xi(m) = r\} = \sum_{k=r}^{m} (-1)^{k-r} \binom{k}{r} B_k(m), \qquad r = 0, 1, \ldots, m.$$

In particular, if $p_j = \delta^j$, $0 < \delta \leq 1$, then formula (6.11) gives the following expressions for the binomial moments of $\xi(m)$:

$$(6.12) \quad B_k(m, \delta) = \frac{1}{T_m} \sum_{l=k}^{m} \binom{m}{l} T_{m-l} \, \sigma(l, k) \, \sigma^l, \quad k = 0, 1, \ldots, m,$$

where $\sigma(l, k)$ are Stirling's numbers of the second kind. Hence the mean of $\xi(m)$ is

$$(6.13) \qquad \mathbf{E}\,\xi(m) = \frac{1}{T_m} \sum_{j=1}^{m} \binom{m}{j} T_{m-j} \, \delta^j.$$

Recalling Dobinski's formula for Bell's numbers:

$$(6.14) \qquad T_m = \frac{1}{e} \sum_{k=0}^{\infty} \frac{k^m}{k!},$$

we eventually obtain

$$(6.15) \qquad \mathbf{E}\,\xi(m) = \frac{1}{e\,T_m} \sum_{k=0}^{\infty} \frac{(k+\delta)^m}{k!} - 1.$$

If $p_j = \delta^j$ then the binomial moments of the random variable η_m can be computed by the formula

$$(6.16) \quad \bar{B}_k(m, \delta) = T_m^{-1} \binom{m}{k} \sum_{l=k}^{m} \binom{m-k}{l-k} T_{m-l} \, \delta^l$$

$$\times \sum_{j=0}^{k} \delta(k, j) \, j^{l-k}, \qquad k = 0, 1, \ldots, m.$$

Hence, applying the inversion formula, we find the exact distribution of η_m:

$$(6.17) \quad \mathbf{P}\{\eta_m = j\} = \sum_{k=j}^{m} (-1)^{k-j} \binom{k}{j} \bar{B}_k(m, \delta), \qquad j = 0, 1, \ldots, m.$$

Let us take $p_j = \lambda_j / m$, $j = 1, 2, \dots, m$. In this case the binomial moments of $\xi(m)$ can be calculated by the formula

(6.18) $$B_k(m) = \frac{\lambda^k}{m^k T_m} \binom{m}{k} \sum_{l=0}^{m-k} \binom{m-k}{l} k^l T_{m-k-l}$$

for $k = 0, 1, \dots, m$.

2. We now put $\Lambda_j = \mathbb{N}_0 = \{0, 1, \dots\}$, $j \in \mathbf{A} \subseteq \mathbb{N} = \{1, 2, \dots\}$. The sequence of generating functions $f_m(x_1, x_2, \dots, x_m)$, $m = 0, 1, 2, \dots$, corresponding to this case satisfies the relationship

(6.19) $$\sum_{m=0}^{\infty} T_m^A \, f_m(x_1, x_2, \dots, x_m; \mathbf{A}) \, \frac{t^m}{m!}$$

$$= \exp\left\{ \sum_{j \in A} \frac{t^j}{j!} + \sum_{j \in A} p_j (x_j - 1) \frac{t^j}{j!} \right\}.$$

In particular, if $f_m(x_{i_1}, x_{i_2}, \dots, x_{i_k})$ is the generating function of the random vector $\left(\xi^{(i_1)}, \xi^{(i_2)}, \dots, \xi^{(i_k)} \right)$ then

$$\sum_{m=0}^{\infty} T_m^A \, f_m(x_{i_1}, x_{i_2}, \dots, x_{i_k}; \mathbf{A}) \, \frac{t^m}{m!}$$

$$= \exp\left\{ \sum_{j \in A} \frac{t^j}{j!} + \sum_{l=1}^{k} p_{i_l} (x_{i_l} - 1) \frac{t^{i_l}}{i_l!} \right\}.$$

Denoting by $f_m(x, \mathbf{A})$ the generating function of the random variable $\xi^{(l)}$ we conclude from the preceding equality that

(6.20) $$\sum_{m=0}^{\infty} T_m^A \, f_m^{(l)}(x; \mathbf{A}) \, \frac{t^m}{m!} = \exp\left\{ p_l (x - 1) \frac{t^l}{l!} + \sum_{j \in A} \frac{t^j}{j!} \right\}.$$

From formula (0.21) of Chapter 3 we know that

$$\sum_{m=0}^{\infty} T_m^A \frac{t^m}{m!} = \exp\left\{ \sum_{j \in A} \frac{t^j}{j!} \right\}.$$

With this relation established, one can find the binomial moments of $\xi^{(l)}$:

(6.21) $$B_{kl}(m) = \frac{(m)_{kl} \, p_k^l}{(l!)^k \, k!} \cdot \frac{T_{m-kl}^A}{T_m^A},$$

where T_m^A is the total number of corresponding partitions of an m-element set. The exact distribution of $\xi^{(l)}$ is as follows:

$$\mathbf{P}\{\xi^{(l)} = r\} = \sum_{k=r}^{[m/l]} (-1)^{k-r} \binom{k}{r} \frac{(m)_{kl} \, p_k^l}{(l!)^k \, k!} \cdot \frac{T_{m-kl}^A}{T_m^A},$$

$$r = 0, 1, \dots, [m/l].$$

4.7 Limiting distributions for partitions with labeled blocks

4.7.1 Basic lemma

First we formulate a lemma which plays an important role in finding the limiting distributions of the random variables considered in the preceding section.

Let r be the unique root of the equation $r \, e^r = m$ and let W_1 and W_2 be arbitrary bounded intervals on the real line, each of which contains the origin.

Lemma 1 *Let*

$$S_m(\alpha, \beta) := \frac{1}{e} \sum_{j=0}^{\infty} \frac{(j+\alpha)^{m-\beta}}{j!}.$$

The asymptotic formula

$$(7.1) \quad S_m(\alpha, \beta) = \frac{1}{\sqrt{r}} \, \exp\left\{ m\left(r + \frac{1}{r} - 1\right) + r\,(\alpha - \beta) - 1 \right\} (1 + o(1))$$

is valid as $m \to \infty$, where $o(1) \to 0$ uniformly for all $\alpha \in W_1$ and $\beta \in W_2$. Also,

$$(7.2) \quad \lim_{m \to \infty} e^{-r\,(\alpha-\beta)} \, \frac{S_m(\alpha, \beta)}{S_m(0,0)} = 1.$$

Proof We take a positive constant A and present $S_m(\alpha, \beta)$ as a sum of three summands:

$$S_m(\alpha, \beta) = S_1 + S_2 + S_3 = \sum_{j=0}^{\theta_1-1} + \sum_{j=\theta_1}^{\theta_2} + \sum_{j=\theta_2+1}^{\infty} \frac{(j+\alpha)^{m-\beta}}{e\,j!},$$

where $\theta_1 := \left[e^r - A \sqrt{m} \right]$, $\theta_2 := \left[e^r + A \sqrt{m} \right]$ and $[x]$ is the integral part of x. One can show that

$$S_2 = \frac{1}{e \sqrt{2\pi r}} \exp \left\{ m \left(r + \frac{1}{r} - 1 \right) + r \left(\alpha - \beta \right) \right\}$$

$$\times \sum_{j=-A \sqrt{m}}^{A \sqrt{m}} \exp \left\{ -\frac{j^2 r^2}{2m} \right\} \frac{r}{\sqrt{m}} \left(1 + o(1) \right).$$

Moreover,

$$S_1 / S_2 = O \left(e^{-C r^2} \right), \qquad S_3 / S_2 = O \left(e^{-C r^2} \right), \qquad C > 0,$$

and all the estimates are uniform for $\alpha \in W_1$ and $\beta \in W_2$. Using the corresponding integral approximation for the sum S_2 we obtain the desired result. $\qquad\qquad\square$

Setting $\alpha = \beta = 0$ we deduce from Lemma 1 the asymptotic representation of Bell's numbers given in [111]:

$$T_m = S_m(0,0) = \frac{1}{\sqrt{r}} \exp \left\{ m \left(r + \frac{1}{r} - 1 \right) - 1 \right\} \left(1 + o(1) \right).$$

Equality (7.2) is an easy consequence of the relation

$$\lim_{m \to \infty} \frac{S_m(\alpha, \beta)}{S_m(0,0)} = \lim_{m \to \infty} \frac{S_2}{S_m(0,0)}.$$

4.7.2 Discrete limiting distributions

Let p_j be the probability of a block of size j being labeled, let $\lambda > 0$ be a constant and let r be the unique positive root of the equation $r e^r = m$.

Theorem 7.1 *If*

$$\lim_{m \to \infty} p_j \left(\frac{\lambda}{r} \right)^{-j} = 1, \qquad j = 1, 2, \dots,$$

then the distribution of the random variable $\xi(m)$ converges, as $m \to \infty$, to a Poisson distribution with parameter $e^\lambda - 1$.

Proof It follows from (6.11) that, by the conditions of the theorem, the binomial moments of $\xi(m)$ can be represented as

$$B_k(m) = \frac{1}{T_m} \sum_{l=0}^{m} \binom{m}{l} T_{m-l} \, \sigma(l, k) \left(\frac{\lambda}{r} \right)^l \left(1 + o(1) \right),$$

where, for Stirling's numbers of the second kind $\sigma(l, k)$, we retain the usual agreement: $\sigma(l, k) = 0$ if $l < k$. Using representation (6.14) for Bell's numbers and the identity

$$k!\, \sigma(l, k) = \sum_{\nu=0}^{k} (-1)^{\nu} \binom{k}{\nu} (k - \nu)^{l},$$

we find

$$B_k(m) = \frac{1}{k!\, T_m} \sum_{\nu=0}^{k} (-1)^{\nu} \binom{k}{\nu} \cdot \frac{1}{e} \sum_{s=0}^{\infty} \frac{\left[s + \lambda\,(k - \nu)/r \right]^{m}}{s!} \left(1 + o(1) \right).$$

Finally, applying Lemma 1 with $\alpha = \lambda\,(k - \nu)/r$ and $\beta = 0$ we deduce that

$$\lim_{m \to \infty} B_k(m) = \frac{1}{k!} (e^{\lambda} - 1)^{k}, \qquad k = 0, 1, \ldots.$$

This completes the proof of the theorem since the sequence $(e^{\lambda} - 1)^{k}/k!$, $k = 0, 1, 2 \ldots$, is the sequence of binomial moments of the Poisson distribution with parameter $e^{\lambda} - 1$. $\square$

As a particular case of the asymptotic formula for the binomial moments of $\xi(m)$ we get

$$\lim_{m \to \infty} \mathbf{E}\, \xi(m) = e^{\lambda} - 1.$$

Corollary *If* $p_j = \delta^j$, $j = 1, 2, \ldots$, $\delta = \lambda/r$ *and* $r\, e^r = m$ *then, as* $m \to \infty$, *the distribution of the random variable* $\xi(m)$ *converges to a Poisson distribution with parameter* $e^{\lambda} - 1$.

Applying the corollary to the bicycle race problem we see that if the probability of the failure-free performance of a bicycle is $\delta = \lambda/r$ with $r \approx \log m$ then, for large m, the average number of repair cars needed is approximately $m\,(\log m)^{-1} - e^{\lambda} + 1$.

Theorem 7.2 *Let* $p_j = \delta^j$, $j = 1, 2, \ldots$, *and* $\delta = \delta(m)$ *vary with* m *in such a way that* $r\delta \to \lambda \in (0, \infty)$ *as* $m \to \infty$. *Then, as* $m \to \infty$, *the limiting distribution of the random variable* η_m *is a distribution uniquely determined by its moments, namely,*

$$(7.3) \quad \lim_{m \to \infty} \mathbf{P}\{ \eta_m = k \} = \frac{\lambda^k}{k!}\, T_k \exp\left\{ -(e^{\lambda} - 1) \right\}, \qquad k = 0, 1, \ldots,$$

where the T_k *are Bell's numbers.*

Proof Under the conditions of the theorem, equality (6.16) takes the form

$$\bar{B}_k(m,\,\delta) = \frac{\delta^k}{T_m}\binom{m}{k}\sum_{j=0}^{k}\sigma(k,\,j)\cdot\frac{1}{e}\sum_{s=0}^{\infty}\frac{(s+j\,\delta)^{m-k}}{s!}\,.$$

Hence, applying Lemma 1 with $\alpha = j\,\delta$ and $\beta = k$ we deduce that

$$\bar{B}_k(m,\,\delta) = \frac{1}{k!}\left(\frac{m\,\delta}{e^\tau}\right)^k\sum_{j=0}^{k}\sigma(k,\,j)\,e^{j\tau\delta}\left(1+o(1)\right).$$

From the equality it follows that

$$\lim_{m\to\infty}\bar{B}_k(m,\,\delta) = \frac{\lambda^k}{k!}\sum_{j=0}^{k}\sigma(k,\,j)\,e^{j\lambda}, \qquad k = 0,\,1,\,\dots\,.$$

The generating function of the binomial moments of the limiting distribution is of the form

$$g(t) = \exp\left\{e^{\lambda\,(t+1)} - e^\lambda\right\}.$$

The distribution can be uniquely recovered from its moments and thus the preceding formula gives the generating function of the limiting distribution

$$f(t) = \exp\left\{e^{\lambda t} - e^\lambda\right\}.$$

Calculating the coefficient of t^k in the expansion of the function as a Maclaurin series, we see that the statement of the theorem is valid. $\quad\square$

It follows immediately from the asymptotic expansions for the binomial moments of η_m that the mean and variance of the limiting distribution are λe^λ and $(\lambda+1)\lambda e^\lambda$ respectively.

Theorem 7.3 *Let* $p_j = \lambda\,j\,m^{-1}$, $j = 1, 2, \dots, m$, *where* $\lambda > 0$ *is a constant. Then the limiting distribution of the random variable* $\xi(m)$ *is Poisson with parameter* λ *as* $m \to \infty$.

Proof From (6.18) and (6.14) we know that

$$B_k(m) = \frac{(m)_k\,\lambda^k}{k!\,m^k\,T_m}\cdot\frac{1}{e}\sum_{s=0}^{\infty}\frac{(s+k)^{m-k}}{s!}, \qquad k = 0,\,1,\,\dots\,.$$

We now apply Lemma 1 with $\alpha = \beta = k$ and obtain

$$\lim_{m\to\infty}B_k(m) = \frac{\lambda^k}{k!}, \qquad k = 0,\,1,\,\dots,$$

which completes the proof of the theorem. $\quad\square$

Theorem 7.4 *If*

$$\lim_{m\to\infty} p_j \left(\frac{\lambda}{r}\right)^j = 1, \qquad\qquad j = 1, 2, \ldots,$$

where $r\,e^r = m$ and $\lambda > 0$ is a constant, then, as $m \to \infty$, for any fixed l, the limiting distribution of the random variable $\xi^{(l)}$ is Poisson with parameter $\lambda^l/l!$.

Proof Using formula (6.21) for the binomial moments of $\xi^{(l)}$, applying Lemma 1 for $\alpha = 0$ and $\beta = kl$, and taking the equality $r\,e^r = m$ into account, we deduce that

$$\lim_{m\to\infty} B_{kl}(m) = \frac{1}{k!}\left(\frac{\lambda^l}{l!}\right)^k, \qquad\qquad k = 0, 1, \ldots.$$

This proves Theorem 7.4.   □

4.7.3 Convergence to a normal distribution

Let $p_j = \delta^j$, $j = 1, 2, \ldots$, where $\delta \in (0, 1)$. In this case the generating function $f_m(x)$ of the random variable $\xi(m)$ takes the form

$$(7.4)\qquad \sum_{m=0}^{\infty} T_m\, f_m(x)\,\frac{t^m}{m!} = \exp\left\{e^t - 1 + (x - 1)\left(e^{\delta t} - 1\right)\right\}.$$

Lemma 2 $m \to \infty$, *the generating function $f_m(x)$ has the asymptotic representation*

$$(7.5)\qquad\qquad f_m(x) = e^{h_m(x)}\left(1 + o(1)\right),$$

where

$$h_m(x) = m\left(\log r - \log R\right) + e^R - e^r + (x - 1)\left(e^{\delta R} - 1\right),$$

$R = R(x)$ *is the unique real solution of the equation*

$$R\,e^R + (x - 1)\,\delta\,R\,e^{\delta R} = m,$$

$r = R(1)$, $r\,e^r = m$ *and $o(1) \to 0$ uniformly for all $x \in W = [1 - x_0, 1 + x_0]$, $0 < x_0 < 1$.*

Proof Applying the method used in the proof of Lemma 1 from Section 2 we deduce the asymptotic representation

$$f_m(x) = \frac{m! \, \exp\left\{ e^R - 1 + (x-1)\left(e^{\delta R} - 1 \right) \right\} \left(1 + o(1) \right)}{T_m \, R^m \, \sqrt{2\pi} \, \left[R(R+1)\, e^R + (x-1)\, \delta R(\delta R + 1)\, e^{\delta R} \right]^{1/2}},$$

where $o(1) \to 0$ uniformly for $x \in W$. Further, using the asymptotic formula for Bell's numbers and taking into account the equality

$$\frac{r(r+1)\, e^r}{R(R+1)\, e^R + (x-1)\, \delta R(\delta R + 1)\, e^{\delta R}} = 1 + o(1),$$

where $o(1) \to 0$ uniformly for $x \in W$, we obtain

$$f_m(x) = \left(\frac{r}{R} \right)^m \exp\left\{ e^R - e^r + (x-1)\left(e^{\delta R} - 1 \right) \right\} \left(1 + o(1) \right).$$

Hence Lemma 2 follows. □

Theorem 7.5 *Let* $p_j = \delta^j$, $j = 1, 2, \ldots$; $0 < \delta < 1$. *Then the distribution of the random variable* $\left(\xi(m) - e^{\delta r} \right) e^{-\delta r/2}$, *where* $r e^r = m$, *converges to the standard normal distribution as* $m \to \infty$.

Proof Equality (7.5) makes it possible to apply Theorem 4.2 of Chapter 1 to prove the desired statement. Indeed, it is straightforward to verify that

$$h'_m(1) = e^{\delta r} - 1, \qquad h''_m(1) = -\frac{r\delta^2}{r+1}\, e^{(2\delta-1)r}$$

and

$$h'''_m(x) = O\left(e^{(3\delta-2)r} \right)$$

uniformly for all $x \in W = [1 - x_0, 1 + x_0]$, $0 < x_0 < 1$. Hence it follows that, uniformly for $x \in W$,

$$\frac{h'''_m(x)}{(h'_m(1) + h''_m(1))^{3/2}} \longrightarrow 0$$

as $m \to \infty$. Now Theorem 4.2 of Chapter 1 allows us to state that the distribution of the random variable

$$\xi'(m) = \frac{\xi(m) - h'_m(1)}{\sqrt{h'_m(1) + h''_m(1)}}$$

is asymptotically normal with parameters $(0, 1)$. □

One can prove that, as $m \to \infty$, the mean and variance of $\xi(m)$ satisfy the relations

$$\mathbf{E}\,\xi(m) = e^{\delta r},$$

$$\operatorname{Var}\xi(m) = e^{\delta r} - \frac{r\,\delta^2}{r+1}\,e^{(2\delta-1)r} + O\,(1),$$

and thus the scaling parameters of Theorem 7.5 are asymptotically equivalent to the mean and variance of $\xi(m)$.

Now we proceed to study the asymptotic distribution of the total number η_m of elements belonging to labeled blocks, assuming that a k-block is labeled with probability δ^k, $0 < \delta < 1$. Under this condition, denoting by $\varphi_m(x)$ the generating function of the random variable η_m, we derive from (6.5) and (6.7) that

$$(7.6) \qquad \sum_{m=0}^{\infty} T_m\,\varphi_m(x)\,\frac{t^m}{m!} = \exp\left\{ e^t - 1 + e^{\delta t x} - e^{\delta t} \right\}.$$

Lemma 3 $m \to \infty$, *the generating function of* $\varphi_m(x)$ *has the asymptotic representation*

$$(7.7) \qquad \varphi_m(x) = e^{\,J_m(x)}\left(1 + o(1)\right),$$

where

$$(7.8) \qquad J_m(x) = m\left(\log r - \log \widetilde{R}\right) + e^{\widetilde{R}} + e^r + e^{x\delta\widetilde{R}} - e^{\delta\widetilde{R}},$$

$\widetilde{R} = \widetilde{R}(x)$ *is the unique real root of the equation*

$$(7.9) \qquad \widetilde{R}\,e^{\widetilde{R}} + x\,\delta\,\widetilde{R}\,e^{x\delta\widetilde{R}} - \delta\,\widetilde{R}\,e^{\delta\widetilde{R}} = m$$

and $r = R(1)$, $r\,e^r = m$ *and* $o(1) \to 0$ *uniformly for* $x \in W = [1 - x_0, 1 + x_0]$, $0 < x_0 < 1$.

Proof From (7.6), using Cauchy's formula, we obtain the asymptotic representation

$$\varphi_m(x) = \frac{m!\,\exp\left\{ e^{\widetilde{R}} - 1 + e^{x\delta\widetilde{R}} - e^{\delta\widetilde{R}} \right\}(1 + o(1))}{\sqrt{2\pi}\,T_m\,\widetilde{R}^m\left(\widetilde{R}^2\,e^{\widetilde{R}} + (x\,\delta\,\widetilde{R})^2\,e^{x\delta\widetilde{R}} - (\delta\,\widetilde{R})^2\,e^{\delta\widetilde{R}} \right)^{1/2}}\,,$$

where $o(1) \to 0$ uniformly for $x \in W$. The remaining part of the proof is similar to that of Lemma 2. $\square$

Theorem 7.6 *Let* $p_j = \delta^j$, $j = 1, 2, \ldots$; $0 < \delta < 1$. *Then the distribution of the random variable*

$$\eta'_m = \left(\eta_m - \delta r e^{\delta r} \right) \left(\delta r e^{\delta r/2} \right)^{-1}, \qquad r e^r = m,$$

converges to the standard normal distribution as $m \to \infty$.

Proof Using definition (7.8) of $J_m(x)$ and equation (7.9) for $\widetilde{R}$ we derive the estimates

$$J'_m(1) = \delta r e^{\delta r},$$

$$J''_m(1) = (\delta r)^2 e^{\delta r} - \frac{\delta^2 r (\delta r + 1)}{r + 1} e^{-(1-2\delta) r},$$

$$J'''_m(x) = O\left(\widetilde{R}^3 e^{\delta \widetilde{R}} \right).$$

Note that the third estimate is uniform for $x \in W$. Applying Lemma 3 and Theorem 4.2 of Chapter 1 we establish the assertion of Theorem 7.6. □

In concluding this chapter we consider the distribution of the random variable $\xi^{(l)}$, the number of labeled blocks of size l in a random partition, under the condition $p_l r^l \to \infty$ as $m \to \infty$.

Theorem 7.7 *Let* l *be a fixed positive integer and* $r e^r = m$. *If* $p_l r^l \to \infty$ *as* $m \to \infty$ *then the distribution of the random variable*

$$\frac{\xi^{(l)} - p_l r^l / l!}{\sqrt{p_l r^l / l!}}$$

converges to the standard normal distribution.

Proof Let $f_m^{(l)}(x)$ be the generating function of the random variable $\xi(l)$. Then, according to (6.5) and (6.8), we have

$$\sum_{m=0}^{\infty} T_m f^{(l)}{}_m(x) \frac{t^m}{m!} = \exp\left\{ e^t - 1 + p_l (x - 1) \frac{t^l}{l!} \right\}.$$

From the equality, by the method used in the proof of Lemma 2, one can obtain the representation

$$f_m^{(l)}(x) = e^{J_m(x)} \left(1 + o(1) \right),$$

where

$$J_m(x) := e^{\bar{R}} - e^r + m \left(\log r - \log \bar{R} \right) + p_l (x - 1) \bar{R} / l!,$$

$$\bar{R}\, e^{\bar{R}} = m - p_l\, \frac{(x-1)\, \bar{R}^l}{(l-1)!}\,,$$

$\bar{R} = \bar{R}(x)$, $r = \bar{R}(1)$ and $o(1) \to 0$ uniformly in a neighborhood W of the point $x = 1$. Further, it is easy to verify that

$$J_m'(1) \;=\; \frac{p_l\, r^l}{l!}\,, \qquad\qquad J_m''(1) \;=\; -\,\frac{r}{e^r\,(r+1)}\,\left(\frac{p_l\, r^{l-1}}{(l-1)!}\right)^2$$

and, in addition, that

$$J_m'''(x) \;=\; O\!\left(\frac{\bar{R}^{3\,(l-1)}}{e^{2\,\bar{R}}}\right)$$

uniformly for $x \in W$. By applying Theorem 4.2 from Chapter 1 we complete the proof of Theorem 7.7.   □

4.7.4 The average number of servicing cars in a bicycle race

We now return to the bicycle race problem. Let us assume that the block of a random partition corresponding to a group of racers is labeled if there is no broken bicycle in the group. Hence it follows that the number of servicing cars is $\zeta(m) = \xi_m - \xi(m)$, where ξ_m is the number of blocks in a random partition of an m-element set and $\xi(m)$ is the number of labeled blocks.

If δ is the probability that any bicycle is in running order then, provided that the serviceability of one bicycle is independent of the other bicycles, the probability that there are no broken bicycles in a group of k racers is δ^k. Therefore, we consider the case where blocks are labeled with probability $p_k = \delta^k$. Using (6.15) and (1.4) we conclude that the average number of servicing cars is

$$\mathbf{E}\,\zeta(m) \;=\; \frac{1}{e\,T_m}\,\sum_{k=0}^{\infty}\frac{(k+1)^m - (k+\delta)^m}{k!}\,.$$

Take $\delta = 1 - v\,m^{-1}$, where $v = o(m/r)$ and $r\,e^r = m$. This means that the serviceability of each bicycle tends to 1 faster than the inverse quantity of the average number of elements in a block of a random partition. Adopting arguments similar to those used in the derivation of equality (7.2) one can show that

$$\mathbf{E}\,\zeta(m) = \left[e^r - e^{r\,(1-v/m)}\right](1 + o(1))\,.$$

Hence we obtain the final asymptotic formula for the average number of servicing cars:

$$\mathbf{E}\,\zeta(m) = v\big(1 + o(1)\big)\,,$$

which is valid for $v = o(m/r)$.

It is not difficult to see that the case under consideration corresponds to the situation where the probability of failure of a bicycle is of the order $o\big((\log m)^{-1}\big)$.

CHAPTER 5

Random Permutations

5.0 Introduction

5.0.1 Λ-permutations

We consider the symmetric group $\mathbb{S}_n$, the set of all permutations of order n. A permutation $s \in \mathbb{S}_n$ is said to belong to a *cyclic class* $\{1^{\alpha_1} 2^{\alpha_2} \ldots n^{\alpha_n}\}$, $1\,\alpha_1 + 2\,\alpha_2 + \cdots + n\,\alpha_n = n$, if it has α_j cycles of length j, $j = 1, 2, \ldots, n$. A permutation s from the given cyclic class is called a Λ-*permutation* if $\alpha_j \in \Lambda_j$, $j = 1, 2, \ldots, n$, where Λ_j is a sequence of numbers whose elements belong to the set $\mathbb{N}_0 = \{0, 1, 2, \ldots\}$ and $\Lambda = (\Lambda_1, \Lambda_2, \ldots)$.

Let $C_n(\Lambda)$ be the number of Λ-permutations of order n, let $C_{nk}(\Lambda)$ be the number of Λ-permutations with k cycles, and let $C_{nk}^l(\Lambda)$ be, for a fixed $1 \le r \le n$, the number of Λ-permutations with k_j cycles of length l_j, where $k_j \in \Lambda_{l_j}$, $j = 1, 2, \ldots, r$, and where the notations $\mathbf{l} := (l_1, \ldots, l_r)$ and $\mathbf{k} := (k_1, \ldots, k_r)$ are used. For the reader's convenience we set

$$C_0(\Lambda) = C_{00}(\Lambda) = C_{00}^0(\Lambda) = 1,$$

where $\mathbf{0}$ is the zero vector of the corresponding dimension. The number of permutations in the cyclic class $\{1^{\alpha_1} 2^{\alpha_2} \ldots n^{\alpha_n}\}$, $1\,\alpha_1 + 2\,\alpha_2 + \cdots + n\,\alpha_n = n$, is calculated by the formula

$$(0.1) \qquad C(\alpha_1, \alpha_2, \ldots, \alpha_n) := n! \prod_{j=1}^{n} \frac{1}{j^{\alpha_j} \, \alpha_j!} \, .$$

The *cyclic indicator* of Λ-permutations is the function

$$(0.2) \quad C_n(x_1, \ldots, x_n; \Lambda) := \sum_{\substack{1\,\alpha_1 + \cdots + n\,\alpha_n = n \\ \alpha_j \in \Lambda_j}} C(\alpha_1, \ldots, \alpha_n) \, x_1^{\alpha_1} \cdots x_n^{\alpha_n},$$

where summation is taken over all the partitions of n in which the number j occurs α_j times and $\alpha_j \in \Lambda_j$, $j = 1, 2, \ldots, n$. From the definition of the cyclic indicator of Λ-permutations it follows that

$$(0.3) \qquad\qquad C_n(\Lambda) = C_n(1, \ldots, 1; \Lambda),$$

$$(0.4) \qquad C_n(x; \Lambda) = \sum_{k=0}^{n} C_{nk}(\Lambda)\, x^k = C_n(x, \ldots, x; \Lambda),$$

$$(0.5) \qquad C_n^{\mathbf{l}}(x_1, \ldots, x_r; \Lambda) = \sum_{\mathbf{k}} C_{n\mathbf{k}}^{\mathbf{l}}(\Lambda)\, x_1^{k_1} \cdots x_r^{k_r}$$

$$= C_n(1, \ldots, x_1, \ldots, x_r, \ldots, 1; \Lambda),$$

where the variables $x_1, \ldots, x_r$ are located at positions $l_1, \ldots, l_r$, respectively, while the remaining coordinates are taken to be 1.

The generating function of the cyclic indicator of Λ-permutations has the form

$$(0.6) \qquad \sum_{n=0}^{\infty} C_n(x_1, \ldots, x_n; \Lambda)\, \frac{t^n}{n!} = \prod_{j=1}^{\infty} \sum_{\alpha_j \in \Lambda_j} \left(\frac{x_j\, t^j}{j} \right)^{\alpha_j} \frac{1}{\alpha_j!}.$$

Setting $C_{nk} := C_{nk}(\mathbf{N}_0)$ and $C_{n\mathbf{k}}^{\mathbf{l}} := C_{n\mathbf{k}}^{\mathbf{l}}(\mathbf{N}_0)$, where $\mathbf{N}_0 := (\mathbb{N}_0, \mathbb{N}_0, \ldots)$, we find from representation (0.6) that

$$(0.7) \qquad C_n(x) := \sum_{k=0}^{n} C_{nk}\, x^k = x\,(x+1) \cdots (x+n-1),$$

$$(0.8) \quad C_n^{\mathbf{l}}(x_1, \ldots, x_r) := \sum_{\mathbf{k}} C_{n\mathbf{k}}^{\mathbf{l}}\, x_1^{k_1} \cdots x_r^{k_r}$$

$$= n! \sum_{\mathbf{k} \in D(n,\mathbf{l})} \frac{(x_1-1)^{k_1}}{l_1^{k_1}\, k_1!} \cdots \frac{(x_r-1)^{k_r}}{l_r^{k_r}\, k_r!},$$

where $D(n,\mathbf{l}) := \{ \mathbf{k} = (k_1, \ldots, k_r) \geq \mathbf{0}\colon k_1 l_1 + \cdots + k_r l_r \leq n \}$. From (0.7) we see that

$$(0.9) \qquad\qquad C_{nk} = |s(n,k)|,$$

where $s(n,k)$ are Stirling's numbers of the second kind determined by the relation

$$x\,(x-1) \cdots (x-n+1) = \sum_{k=0}^{n} s(n,k)\, x^k.$$

Formula (0.8) implies

$$(0.10) \qquad C_{nk}^1 = n! \prod_{j=1}^{r} \frac{1}{l_j^{k_j} k_j!} \sum_{(j_1,\ldots,j_r)} \frac{(-1)^{j_1+\cdots+j_r}}{l_1^{j_1}\cdots l_r^{j_r}\, j_1!\cdots j_r!},$$

where summation is taken over all the tuples $(j_1,\ldots,j_r)$ satisfying the conditions

$$j_1 \geq 0,\ldots, j_r \geq 0 \qquad \text{and} \qquad \sum_{i=1}^{r} (k_i + j_i)\, l_i \leq n.$$

Denoting by $C_{nk}^{(l)}$ the number of permutations of order n with k cycles and setting $r = 1$ in (0.10) we obtain

$$(0.11) \qquad C_{nk}^{(l)} = \frac{n!}{l^k\, k!} \sum_{j=0}^{[n/l]-k} \frac{(-1)^j}{l^j\, j!}, \qquad\qquad k = 0, 1, \ldots, [n/l].$$

5.0.2 A-permutations

Let A be a sequence of positive integers. A permutation s of order n is called an *A-permutation* if the length of each of its cycles is a member of the sequence A.

Let C_n^A be the total number of A-permutations of order n, let C_{nk}^A be the number of A-permutations with k cycles and let C_{nk}^{1A} be the number of A-permutations with k_j cycles of length l_j, where $l_j \in A$, $j = 1, 2, \ldots, r$, $1 = (l_1,\ldots,l_r)$ and $\mathbf{k} = (k_1,\ldots,k_r)$. The corresponding generating functions are

$$(0.12) \qquad\qquad \sum_{n=0}^{\infty} C_n^A \frac{t^n}{n!} = e^{a(t)},$$

$$(0.13) \qquad\qquad \sum_{n=0}^{\infty} \sum_{k=0}^{n} C_{nk}^A \frac{t^n}{n!}\, x^k = e^{x\, a(t)},$$

$$(0.14) \quad \sum_{n=0}^{\infty} \sum_{\mathbf{k}} C_{nk}^{1A} \frac{t^n}{n!}\, x_1^{k_1} \cdots x_r^{k_r} = \exp\left\{ a(t) + \sum_{j=1}^{r} (x_j - 1) \frac{t^{l_j}}{l_j} \right\},$$

where

$$(0.15) \qquad\qquad a(t) := \sum_{j \in A} \frac{t^j}{j}.$$

Denote by E_{nk}^A the number of permutations $s \in \mathbb{S}_n$ with exactly k cycles whose lengths are elements of the specified sequence A. The generating function of the numbers E_{nk}^A, $k \le n$, $n = 0, 1, \ldots$, is given by the formula

$$(0.16) \qquad \sum_{n=0}^{\infty} \sum_{k=0}^{n} E_{nk}^A \frac{t^n}{n!} \, x^k = \frac{1}{1-t} \, \exp\left\{ (x-1) \, a(t) \right\},$$

where $a(t)$ is the same as in (0.15).

5.0.3   Permutations with congruent cycles

A permutation s of order n is called a *permutation with congruent cycles* if there exist α and β, $1 \le \beta \le \alpha$, such that the length of any cycle of s and the number β are congruent modulo α. We denote by $S_n(\alpha, \beta)$ the set of all such permutations. It is clear that if $s \in S_n(\alpha, \beta)$ then s is an A-permutation with $A = \{\beta, \, \alpha + \beta, \, 2\alpha + \beta, \, \ldots\}$ and

$$(0.17) \qquad a(t; \alpha, \beta) = \sum_{j=0}^{\infty} \frac{t^{\alpha j + \beta}}{\alpha j + \beta}$$

as the function $a(t)$.

Using the αth roots of unity we can write

$$(0.18) \quad a(t; \alpha, \beta) = -\frac{1}{\alpha} \sum_{k=1}^{\alpha} \exp\left\{ -2\pi\beta k\, i\,/\alpha \right\} \, \log\left(1 - t \exp\left\{ 2\pi k\, i\,/\alpha \right\}\right),$$

where the principal value of the logarithm is taken. Representation (0.18) gives, in particular, that

$$(0.19) \qquad a(t; \alpha, \alpha) = -\frac{1}{\alpha} \, \log\left(1 - t^\alpha\right)$$

and

$$(0.20) \qquad a(t; \alpha, \alpha/2) = \frac{1}{\alpha} \, \log\left(\frac{1 + t^{\alpha/2}}{1 - t^{\alpha/2}}\right).$$

Let $C_{nk}(\alpha, \beta)$ be the number of permutations $s \in S_n(\alpha, \beta)$ with k cycles and let

$$(0.21) \qquad C_n(x; \alpha, \beta) := \sum_{k=0}^{n} C_{nk}(\alpha, \beta) \, x^k.$$

From (0.13), (0.19) and (0.20) it follows that

$$(0.22) \quad C_n(x; \alpha, \alpha) = n! \binom{(x+n)/\alpha - 1}{n/\alpha}, \qquad \alpha \mid n,$$

$$(0.23) \quad C_n(x; \alpha, \alpha/2) = n! \sum_{j=1}^{2n/\alpha} \binom{x/\alpha}{2n/\alpha - j} \binom{x/\alpha + j - 1}{j}, \frac{\alpha}{2} \mid n,$$

and $C_n(x; \alpha, \alpha) = 0$ if α does not divide n; $C_n(x; \alpha, \alpha/2) = 0$ if $\alpha/2$ does not divide n. Since $C_n(\alpha, \beta)$, the number of permutations in $S_n(\alpha, \beta)$, is determined by the relation

$$(0.24) \qquad\qquad C_n(\alpha, \beta) = C_n(1; \alpha, \beta),$$

it is not difficult to deduce from (0.22) and (0.23) the explicit formulae for the quantities $C_n(\alpha, \alpha)$ and $C_n(\alpha, \alpha/2)$.

Let $E_{nk}(\alpha, \beta)$ be the number of permutations of order n which have exactly k cycles whose lengths are congruent modulo α to β and let

$$(0.25) \qquad\qquad E_n(x; \alpha, \beta) := \sum_{k=0}^{n} E_{nk}(\alpha, \beta)\, x^k.$$

From (0.16) we find

$$(0.26) \qquad \sum_{n=0}^{\infty} E_n(x; \alpha, \beta) \frac{t^n}{n!} = \frac{1}{1-t} \exp\{(x-1)\, a(t; \alpha, \beta)\}.$$

With the aid of formula (0.19) we obtain

$$(0.27) \qquad E_n(x; \alpha, \alpha/2) = n! \binom{(x-1)/\alpha + [n/\alpha]}{[n/\alpha]}.$$

Denote by $D(n, \mu, \nu)$ the number of permutations of order n with exactly μ cycles of odd length and exactly ν cycles of even length and put

$$(0.28) \qquad D_n(x, y) := \sum_{\mu=0}^{n} \sum_{\nu=0}^{n} D(n, \mu, \nu)\, x^\mu\, y^\nu.$$

It is straightforward to check that

$$\sum_{n=0}^{\infty} D_n(x, y) \frac{t^n}{n!} = (1+t)^{(x-y)/2} (1-t)^{-(x+y)/2}.$$

Hence we see that

$$(0.29) \qquad D_n(x, y) = n! \sum_{k=0}^{n} \binom{(x-y)/2}{n-k} \binom{(x+y)/2 + k - 1}{k}.$$

From this equality one can find, setting $x = 1$ or $y = 1$, the generating functions for the numbers of permutations with a prescribed number of cycles of even or odd length respectively.

5.0.4 Even and odd permutations

The *decrement* of a permutation $s \in \mathbb{S}_n$ is the difference $n - k$, where k is the number of cycles in s. A permutation s is said to be *even* (*odd*) if its decrement is an even (odd) number. Multiplication of any permutation by a transposition changes the number of its cycles by one, that is, changes the parity of the permutation. Using this property one can establish a bijection between the sets of even and odd permutations. Hence it follows that the number of even permutations of order n coincides with that of odd permutations and is equal to $\frac{1}{2} n!$.

We denote by $C_n^e(x_1, \ldots, x_n)$ and $C_n^o(x_1, \ldots, x_n)$ the cyclic indicators of the even and odd permutations of order n respectively. It follows from (0.2) that

$$(0.30) \qquad 2\,C_n^e(x_1, x_2, \ldots, x_n) = C_n(x_1, x_2, \ldots, x_n)$$

$$+ C_n(x_1, -x_2, x_3, -x_4, \ldots)$$

and

$$(0.31) \qquad 2\,C_n^o(x_1, x_2, \ldots, x_n) = C_n(x_1, x_2, \ldots, x_n)$$

$$- C_n(x_1, -x_2, x_3, -x_4, \ldots)\,.$$

Hence, setting $x_1 = \cdots = x_n = x$, we obtain the generating functions for C_{nk}^e and C_{nk}^o, the numbers of even and odd permutations with k cycles respectively.

5.1 Cycles in random permutations

5.1.1 The number of cycles

We consider the uniform probability distribution on the symmetric group $\mathbb{S}_n$ of all permutations of order n and denote by ξ_n the number of cycles in a random permutation $s \in \mathbb{S}_n$.

According to (0.9) the explicit distribution of ξ is given by Stirling's numbers of the first kind:

$$\mathbf{P}\left\{\xi_n = k\right\} = \frac{|\,s(n, k)\,|}{n!}\,, \qquad\qquad k = 0, 1, \ldots, n\,.$$

Denoting by $P_n(x)$ the generating function of ξ_n and using equality (0.7) we find that

$$(1.1) \qquad P_n(x) = \binom{x + n - 1}{n} = \frac{\Gamma(x + n)}{\Gamma(x)\,\Gamma(n + 1)}\,,$$

where $\Gamma(x)$ is the gamma-function. By differentiating $P_n(x)$ at the point $x = 1$ it is not difficult to show that

$$\mathbf{E}\,\xi_n = \sum_{j=1}^{n} \frac{1}{j}\,, \qquad \mathrm{Var}\,\xi_n = \sum_{j=1}^{n} \frac{1}{j} - \sum_{j=1}^{n} \frac{1}{j^2}\,.$$

Now, as $n \to \infty$, we use the formulae

$$\sum_{j=1}^{n} \frac{1}{j} = \log n + C + o(1)\,, \qquad C = 0.5772\ldots\,,$$

$$\sum_{j=1}^{n} \frac{1}{j^2} = \frac{\pi^2}{6} + o(1)\,,$$

and find the asymptotic expansions

(1.2)
$$\begin{cases} \mathbf{E}\,\xi_n = \log n + C + o(1)\,, \\ \mathrm{Var}\,\xi_n = \log n + C - \tfrac{1}{6}\,\pi^2 + o(1)\,. \end{cases}$$

The two theorems formulated below were first proved by Goncharov in [49] by the method of characteristic functions. Here we prove them by using generating functions and Curtiss' theorem.

Theorem 1.1 *The distribution of the random variable*

$$\eta_n' = (\xi_n - \log n)\,(\log n)^{-1/2}$$

is, as $n \to \infty$, asymptotically normal with parameters $(0,1)$, that is,

$$\lim_{n\to\infty} \mathbf{P}\{\eta_n' < u\} = \frac{1}{\sqrt{2\pi}} \int_{-\infty}^{u} e^{-y^2/2}\,dy\,.$$

Proof From (1.1) we see that

(1.3)
$$P_n(x) = \frac{n^{x-1}}{\Gamma(x)}\left(1 + o(1)\right),$$

where $o(1) \to 0$, as $n \to \infty$, uniformly for all x from a neighborhood $1 - \delta < x < 1 + \delta, 0 < \delta < 1$, of the point $x = 1$. Hence we find the asymptotic representation of the moment generating function of ξ_n:

(1.4)
$$M_n(t) = \frac{n^{e^t-1}}{\Gamma(e^t)}\left(1 + o(1)\right),$$

where $o(1) \to 0$, as $n \to \infty$, uniformly for $t \in (-\delta', \delta')$, $\delta' > 0$.

We put $\sigma^2 = \log n$ and consider the moment generating function of the random variable η_n:

(1.5) $\widetilde{M}_n(t) = \exp\{-t \log n/\sigma\} \, M_n(t/\sigma).$

From equalities (1.4) and (1.5) it follows that

$$\lim_{n\to\infty} \widetilde{M}_n(t) = e^{t^2/2}$$

for each $t \in (-\delta', \delta')$, $\delta' > 0$.

The right-hand side of the preceding equality coincides with the moment generating function of the standard normal distribution. Now the validity of Theorem 1.1 follows from Curtiss' theorem.   □

Denote by $\varkappa_{nl}$ the number of cycles of unit length in a permutation of order n chosen at random and equiprobably from $\mathbb{S}_n$.

Theorem 1.2 *The distribution of the random variable $\varkappa_{nl}$ converges, as $n \to \infty$, to a Poisson distribution with parameter $\lambda = l^{-1}$, that is,*

$$\lim_{n\to\infty} \mathbf{P}\{\varkappa_{nl} = k\} = \frac{1}{l^k \, k!}\, e^{-1/l}, \qquad\qquad k = 0, 1, \dots .$$

Proof From formula (0.8), setting $r = 1$, we find the explicit expression for the generating function of $\varkappa_{nk}$:

$$P_n(x; l) = \sum_{k=0}^{[n/l]} \frac{(x-1)^k}{l^k \, k!}\ .$$

By the equality

$$\lim_{n\to\infty} P_n(x; l) = e^{(x-1)/l},$$

we conclude that, for each $x \in [0, 1]$, the generating function of $\varkappa_{nl}$ converges, as $n \to \infty$, to the generating function of the Poisson distribution with parameter $\lambda = l^{-1}$, which completes the proof of Theorem 1.2.   □

Theorem 1.2 can easily be deduced from the explicit expression for the distribution of $\varkappa_{nk}$:

$$\mathbf{P}\{\varkappa_{nl} = k\} = \frac{1}{l^k \, k!} \sum_{j=1}^{[n/l]-k} \frac{(-1)^j}{l^j \, j!}\ , \qquad\qquad k = 0, 1, \dots, [n/l],$$

obtained by the expansion of $P_n(x; l)$ in powers of x; the same statement also follows from the formula for the binomial moments of $\varkappa_{nk}$:

$$B_{kn}^{(l)} = \begin{cases} (l^k \, k!)^{-1}, & k = 0, 1, \dots, [n/l], \\ 0, & k > [n/l]. \end{cases}$$

Let $l_1 < l_2 < \cdots < l_r$ be a sequence of positive integers and let $\varkappa_{nl_j}$ be the number of cycles of length l_j in a random permutation of order n. By (0.8) the generating function

$$P_n(x_1, x_2, \ldots, x_r) = \sum_{k_1, \ldots, k_r = 0} \mathbf{P}\{\varkappa_{nl_1} = k_1, \ldots, \varkappa_{nl_r} = k_r\} x_1^{k_1} \cdots x_r^{k_r}$$

of the r-dimensional random variable $(\varkappa_{nl_1}, \varkappa_{nl_2}, \ldots, \varkappa_{nl_r})$ is given by the formula

$$P_n(x_1, x_2, \ldots, x_r) = \sum_{\mathbf{k} \in D(n,\,\mathbf{l})} \frac{(x_1 - 1)^{k_1}}{l_1^{k_1} k_1!} \cdots \frac{(x_r - 1)^{k_r}}{l_r^{k_r} k_r!}.$$

Using this expression, or indeed directly relation (0.10) one can establish the following result [66].

Theorem 1.3 *For any fixed $l_1, l_2, \ldots, l_r$ and $k_1, k_2, \ldots, k_r$,*

$$\mathbf{P}\{\varkappa_{nl_1} = k_1, \ldots, \varkappa_{nl_r} = k_r\} = \prod_{j=1}^{r} \frac{1}{l_j^{k_j} k_j!} e^{-1/l_j} + o(1), \qquad n \to \infty.$$

Hence it follows that the random variables $\varkappa_{nl_1}, \varkappa_{nl_2}, \ldots, \varkappa_{nl_r}$ are asymptotically independent as $n \to \infty$ and their limiting distributions are Poisson with parameters $1/l_1, 1/l_2, \ldots, 1/l_r$ respectively.

In this connection we note that the random variable ξ_n, the number of cycles in a random equiprobable permutation, can be written as

$$\xi_n = \varkappa_{n1} + \varkappa_{n2} + \cdots + \varkappa_{nn}, \qquad n = 1, 2, \ldots,$$

where the summands are dependent. There exists a representation of ξ_n as a sum of independent random variables [42]. To obtain the representation we define a stochastic procedure which constructs a random permutation s in n steps. At the kth step we choose, at random and equiprobably, the image $s(k)$ of the element k under the permutation s from the set of elements which does not contain the previously chosen elements $s(1), \ldots, s(k-1)$. As a result of this procedure each permutation of order n occurs with probability $(n!)^{-1}$. We associate a random variable η_n with the kth step of the procedure. The random variable η_n is equal to 1 if an element is chosen for $s(k)$ which closes a cycle of s constituted by this element and some previously chosen elements, and is equal to 0 otherwise.

It is clear that

$$\mathbf{P}\{\eta_k = 1\} = \frac{1}{n - k + 1}, \qquad \mathbf{P}\{\eta_k = 0\} = \frac{n - k}{n - k + 1}.$$

Moreover, $\xi_n = \eta_1 + \cdots + \eta_n$ and the random variables $\eta_1, \ldots, \eta_n$ are independent and satisfy Lindeberg's condition. These facts allow us to derive Theorem 1.1 as a corollary of the central limit theorem.

5.1.2 Cycles of even and odd lengths

We denote by $\xi_n^{(1)}$ the number of cycles having odd length and by $\xi_n^{(2)}$ the number of cycles having even length in a random equiprobable permutation. Setting $y = 1$ and $x = 1$, respectively, in (0.29) and dividing both sides of the equalities obtained by $n!$, we find the explicit expressions for the generating functions of the random variables $\xi_n^{(1)}$ and $\xi_n^{(2)}$:

$$P_n^{(1)}(x) = \sum_{k=0}^{n} \binom{(x-1)/2}{n-k}\binom{(x+1)/2+k-1}{k},$$

$$P_n^{(2)}(y) = \sum_{k=0}^{n} \binom{(1-y)/2}{n-k}\binom{(y+1)/2+k-1}{k}.$$

Direct calculations show that, as $n \to \infty$,

$$P_n^{(1)}(x) = \binom{(x+1)/2+n-1}{n}\left(1 + \frac{(x-1)\,R_n^{(1)}(x)}{2}\right),$$

$$P_n^{(2)}(y) = \binom{(y+1)/2+n-1}{n}\left(1 - \frac{(y-1)\,R_n^{(2)}(y)}{2}\right),$$

where

$$R_n^{(1)}(x) = O(1), \qquad R_n^{(2)}(y) = O(1)$$

and the estimates are uniform for $1 - \delta < x, y < 1 + \delta$, $0 < \delta < 1$. Using Stirling's formula we have, as $n \to \infty$,

$$P_n^{(i)}(x) = \frac{n^{(x-1)/2}}{\Gamma((x+1)/2)}\left(1 + (x-1)\,O(1)\right), \qquad i = 1, 2,$$

where $O(1)$ is uniformly bounded for all $x \in [1 - \delta, 1 + \delta]$, $0 < \delta < 1$.

Adopting arguments similar to those used in proving Theorem 1.1 one can obtain the following statement.

Theorem 1.4 *The distributions of the random variables*

$$\tilde{\xi}_n^{(i)}(x) = (\xi_n^{(i)}(x) - \tfrac{1}{2}\log n)(\tfrac{1}{2}\log n)^{-1/2}, \qquad i = 1, 2,$$

converge to the standard normal distribution as $n \to \infty$.

5.1.3 Cycles whose lengths are multiples of a given number

Let $\xi_n^{(\alpha)}$ be the number of cycles in a random equiprobable permutation whose lengths are multiples of the number α. From (0.27) we find the generating function of $\xi_n^{(\alpha)}$:

$$P_n^{(\alpha)}(x) = \left(\frac{(x-1)/\alpha + [n/\alpha]}{[n/\alpha]} \right).$$

Applying Stirling's formula we get, as $n \to \infty$,

$$P_n^{(\alpha)}(x) = \frac{[n/\alpha]^{(x-1)/\alpha}}{\Gamma\big((x-1)/\alpha + 1\big)} \big(1 + o(1)\big),$$

where $o(1) \to 0$ uniformly for x, $1 - \delta < x < 1 + \delta$, $0 < \delta < 1$.

Using this asymptotic representation it is not difficult to prove the following theorem.

Theorem 1.5 *The following two assertions are valid as $n \to \infty$:*
(a) *If*

$$\frac{1}{\alpha} \log\left[\frac{n}{\alpha}\right] \longrightarrow \lambda < \infty$$

then the distribution of $\xi_n^{(\alpha)}$ converges to a Poisson distribution with parameter λ.

(b) *If*

$$\frac{1}{\alpha} \log\left[\frac{n}{\alpha}\right] \longrightarrow \infty$$

then the random variable

$$\eta_n^{(\alpha)} = \left(\xi_n^{(\alpha)}(x) - \frac{1}{\alpha} \log\left[\frac{1}{\alpha}\right] \right) \left(\frac{1}{\alpha} \log\left[\frac{1}{\alpha}\right] \right)^{-1/2}$$

has asymptotically normal distribution with parameters $(0,1)$.

Proof Under condition (a) we have

$$\lim_{n \to \infty} P_n^{(\alpha)}(e^t) = \exp\left\{\lambda\left(e^t - 1\right)\right\}$$

for any $t \in (-\delta', \delta')$, $\delta' > 0$. Hence it follows that the Poisson distribution with parameter λ is limiting for the distribution of the random variable $\xi_n^{(\alpha)}$ as $n \to \infty$.

Item (b) is proved by arguments similar to those used in the proof of Theorem 1.1.   □

5.1.4 Even and odd permutations

We consider the uniform distribution on the set of all even (odd) permutations. Let $\xi_n^{(e)}$ ($\xi_n^{(o)}$) be the number of cycles in an even (odd) permutation chosen at random and equiprobably. According to (0.30) and (0.31) the generating functions of $\xi_n^{(e)}$ and $\xi_n^{(o)}$ have, respectively, the form:

$$P_{n_1}(x) = \binom{x+n-1}{n} + \binom{x}{n},$$

$$P_{n_2}(x) = \binom{x+n-1}{n} - \binom{x}{n}.$$

Hence we find explicit expressions for the distributions of the random variables in question by Stirling's numbers of the first kind:

$$\mathbf{P}\left\{\xi_n^{(e)} = k\right\} = \frac{|s(n,k)| + s(n,k)}{n!},$$

$$\mathbf{P}\left\{\xi_n^{(o)} = k\right\} = \frac{|s(n,k)| - s(n,k)}{n!}.$$

We have the following asymptotic representations for $P_{ni}(x)$ as $n \to \infty$:

$$P_{ni}(x) = \frac{n^{x-1}}{\Gamma(x)} + (1-x)\, R_{ni}(x), \qquad i = 1, 2,$$

where $R_{ni}(x) = O(1)$, $n \to \infty$, uniformly for $x \in (1-\delta, 1+\delta)$, $0 < \delta < 1$.

Using a standard method, one can derive the following statement from the preceding representations.

Theorem 1.6 *The random variable*

$$\eta_n^{(e)} = (\xi_n^{(e)} - \log n)(\log n)^{-1/2}$$

is asymptotically normal with parameters $(0,1)$ as $n \to \infty$. The same statement is valid for the random variable

$$\eta_n^{(o)} = (\xi_n^{(o)} - \log n)(\log n)^{-1/2}.$$

If $\varkappa_{nl}^e$ ($\varkappa_{nl}^o$) is the number of cycles of length l in an even (odd) permutation chosen at random and equiprobably and $n \to \infty$, then the limiting distribution of $\varkappa_{nl}^e$ ($\varkappa_{nl}^o$) is Poisson with parameter $\lambda = l^{-1}$. One can prove this fact, for example, by finding binomial moments of $\varkappa_{nl}^e$ ($\varkappa_{nl}^o$) and analyzing their behavior as $n \to \infty$.

5.2 Ordered sample of a permutation

We consider a permutation s of order n and arrange the sequence of numbers specifying the lengths of cycles of s in nondecreasing order. The resulting sequence is called the *ordered sample* of the permutation s. The minimal and maximal terms of the sequence correspond, respectively, to the minimal and maximal cycle length of s. We have occasion to use the uniform distribution on the set of all permutations of order n. Denote by $\nu_m^{(n)}$ the mth term of the ordered sample of a random permutation; the first and last terms of the ordered sample are denoted by $\nu^{(n)}$ and $\tilde{\nu}^{(n)}$ respectively.

Obviously,

$$\mathbf{P}\{\nu_m^{(n)} \leq k\} = \mathbf{P}\{\varkappa_{1n} + \cdots + \varkappa_{kn} \geq m\}, \qquad k = 1, 2, \ldots, n,$$

where $\varkappa_{ln}$ is the number of cycles of length l in a random permutation. This equality and Theorem 1.3 imply (see [66])

$$\mathbf{P}\{\nu_m^{(n)} \leq k\} = \sum_{j=m}^{\infty} \frac{\lambda_k^j}{j!} e^{-\lambda_k} + o(1), \qquad \lambda_k := \sum_{i=1}^{k} i^{-1},$$

as $n \to \infty$. Hence it follows that

$$\mathbf{P}\{\nu^{(n)} = k\} = (1 - e^{-1/k}) \exp\left\{-\sum_{j=1}^{k-1} j^{-1}\right\} + o(1);$$

here we have assumed $\sum_1^0 = 0$.

We now investigate the distribution function of the random variable $\mu^{(n)} = \tilde{\nu}^{(n)} n^{-1}$:

$$F(n, m) = \mathbf{P}\{\mu^{(n)} < m/n\}, \qquad m = 1, 2, \ldots, n.$$

It is not difficult to check that

$$(2.1) \qquad \sum_{n=0}^{\infty} F(n, m)\, t^n = (1 - t)^{-1} \exp\left\{-\sum_{\nu=m}^{\infty} t^\nu/\nu\right\}.$$

We put

$$\lambda := \left[\frac{n}{m}\right], \qquad \frac{1}{\lambda+1} < \frac{m}{n} \leq \frac{1}{\lambda}, \qquad \lambda \geq 1,$$

and introduce the notations

$$S_0(m, n) := 1, \qquad S_h(m, n) := \sum_{\substack{\nu \geq m \\ \Sigma \nu_i \leq n}} \frac{1}{\nu_1 \nu_2 \cdots \nu_h}, \qquad h \geq 1,$$

setting $S_h(m,n) = 0$ if $h \geq \lambda + 1$. From (2.1) it follows that

$$F(m,n) = \sum_{h=0}^{\lambda} \frac{(-1)^h}{h!} S_h(m,n).$$

For any fixed numbers x' and x'', $(\lambda + 1)^{-1} \leq x' < x'' < \lambda^{-1}$, we have

$$\mathbf{P}\left\{ x' \leq \mu^{(n)} < x'' \right\} = \sum_{h=0}^{\lambda} \frac{(-1)^h}{h!} \left\{ S_h([nx''], n) - S_h([nx'], n) \right\}.$$

If $m, n \to \infty$ in such a way that $x' \leq mn^{-1} \leq x''$ then (see [49])

$$(2.2) \quad \lim_{n \to \infty} \mathbf{P}\left\{ x' \leq \mu^{(n)} < x'' \right\} = \sum_{h=1}^{\lambda} \frac{(-1)^h}{h!} \left\{ J_h(x'', 1) - J_h(x', 1) \right\},$$

where

$$J_0(m,n) := 1, \qquad J_h(m,n) := \int \cdots \int_{\substack{x_i \geq m \\ \Sigma x_i < n}} \frac{dx_1}{x_1} \cdots \frac{dx_h}{x_h}, \qquad h \geq 1.$$

Indeed, $S_h(m,n)$ is a Riemann sum for the integral $J_h(m,n)$, which obeys the homogeneity property: $J_h(km, kn) = J_h(m,n)$ for any $k > 0$. Therefore, $J_h(m,n) = J_h(x,1)$ with $x = mn^{-1}$. Equality (2.2) yields

$$G(x) = \lim_{n \to \infty} \mathbf{P}\left\{ (1+\lambda)^{-1} \leq \mu^{(n)} < x \right\}$$

$$= \sum_{h=1}^{\lambda} \frac{(-1)^h}{h!} \left\{ J_h(x,1) - J_h((\lambda+1)^{-1}, 1) \right\}$$

if

$$\frac{1}{\lambda+1} < x \leq \frac{1}{\lambda}, \qquad\qquad \lambda = 1, 2, \ldots .$$

Via this transform it is not difficult to find the density of the limiting distribution. Indeed, it is easy to show that

$$\frac{d}{dx} J_h(x,1) = -\frac{h}{x} J_{h-1}(x, 1-x).$$

Using this equality we obtain

$$g(x) = G'(x) = \frac{1}{x} \sum_{h=0}^{\lambda-1} \frac{(-1)^h}{h!} J_h(x, 1-x)$$

if

$$\frac{1}{\lambda+1} < x \le \frac{1}{\lambda}, \qquad \lambda = 1, 2, \ldots,$$

and thus the density has distinct analytic representations in intervals of the form $1/2 < x \le 1$, $1/3 < x \le 1/2$, $\ldots$, $(\lambda+1)^{-1} < x \le \lambda^{-1}, \ldots$. Here we give without proof the limiting value of the expectation of $\mu^{(n)}$ from [48]:

$$\lim_{n\to\infty} \mathbf{E}\,\mu^{(n)} = 0.6243 \ldots .$$

We note in conclusion that the limiting distributions of the intermediate and right terms of the ordered sample of a random permutation are known. The reader can find the corresponding results, for instance, in [66].

5.3   Permutations with congruent cycles

We consider a set $S_n(\alpha, \beta)$ of permutations with congruent cycles, that is, the set of such permutations of order n, each cycle of which is congruent modulo α to β. We assign the uniform probability distribution on the set $S_n(\alpha, \beta)$ and consider the random variable $\xi_n(\alpha, \beta)$, the number of cycles in a random permutation $s \in S_n(\alpha, \beta)$. From (0.21) and (0.24) we find the generating function of the random variable $\xi_n(\alpha, \beta)$:

$$(3.1) \qquad P_n\big(x; \alpha, \beta\big) = C_n\big(x; \alpha, \beta\big) \,/\, C_n(\alpha, \beta).$$

We confine ourselves to considering the cases $\beta = \alpha$ and $\beta = \alpha/2$ only (if α is even). Using (0.22) and (0.23) we prove the following lemma giving the asymptotic formulae for $C_n(x; \alpha, \alpha)$ and $C_n(x; \alpha, \alpha/2)$ as $n \to \infty$.

Lemma *If $n \to \infty$ and $\alpha \ge 2$ then*

$$(3.2) \qquad C_n\big(x; \alpha, \alpha\big) = \frac{n!}{\Gamma(x/\alpha)\,(n/\alpha)^{\,1-x/\alpha}} \left(1 + o(1)\right), \quad \alpha \mid n,$$

$$(3.3) \qquad C_n\big(x; \alpha, \alpha/2\big) = \frac{n!\ 2^{2x/\alpha - 1}}{\Gamma(x/\alpha)\,(n/\alpha)^{\,1-x/\alpha}} \left(1 + o(1)\right), \quad \frac{\alpha}{2} \mid n,$$

where $o(1) \to 0$ uniformly for $x \in [1 - \delta, 1 + \delta]$, $0 < \delta < 1$.

Proof Using Stirling's formula for the gamma-function as $k \to \infty$, it is not difficult to obtain the following asymptotic estimates:

$$(3.4) \qquad \binom{y + k - 1}{k} = \frac{1}{\Gamma(y)\,k^{1-y}} \left(1 + o(1)\right),$$

$$(3.5) \qquad \binom{y}{k} = \frac{(-1)^k\,y\,(y - 1)}{\Gamma(2 - y)\,k^{1+y}} \left(1 + o(1)\right),$$

where $o(1) \to 0$ uniformly for $y \in [1 - \delta, 1 + \delta]$, $0 < \delta < 1$.

Now relation (3.2) follows immediately from (3.4) and it thus remains to demonstrate (3.3).

We fix a number $\nu \in (0, \frac{1}{2})$ and, using equality (0.23), write the corresponding sum as

$$(3.6) \qquad \frac{1}{n!}\, C_n(x;\, \alpha,\, \alpha/2) = S_1 + S_2 + S_3$$

$$= \sum_{j=0}^{(n/\alpha)^\nu} + \sum_{j=(n/\alpha)^\nu+1}^{2n/\alpha - (n/\alpha)^\nu - 1} + \sum_{j=2n/\alpha - (n/\alpha)^\nu}^{2n/\alpha}.$$

Applying (3.4) we find

$$S_3 = \frac{1}{\Gamma(x/\alpha)\,(2n/\alpha)^{1-x/\alpha}} \sum_{k=0}^{(n/\alpha)^\nu} \binom{x/\alpha}{k} \left(1 + o(1)\right).$$

Hence it follows that

$$(3.7) \qquad S_3 = \frac{2^{2x/\alpha - 1}}{\Gamma(x/\alpha)\,(n/\alpha)^{1-x/\alpha}} \left(1 + o(1)\right),$$

where, in both preceding equalities, $o(1) \to 0$ uniformly for x from the domain specified by the conditions of the lemma.

Using (3.5) we get

$$S_1 = \frac{(-1)^{2n/\alpha}\,(x/\alpha)\,(x/\alpha - 1)}{\Gamma(2 - x/\alpha)\,(2n/\alpha)^{1+x/\alpha}} \sum_{j=0}^{(n/\alpha)^\nu} \binom{-x/\alpha}{j} \left(1 + o(1)\right),$$

which implies the estimate

$$(3.8) \qquad S_1 = O\left(S_3\, n^{-2x/\alpha}\right),$$

which is uniform for x from the domain indicated above. In turn, we write S_2 as

$$(3.9) \qquad S_2 = S_2^{(1)} + S_2^{(2)} = \sum_{j=(n/\alpha)^\nu+1}^{n/\alpha - 1} + \sum_{j=n/\alpha}^{2n/\alpha - (n/\alpha)^\nu - 1}.$$

Recalling (3.4) and (3.5) we obtain

$$|S_2^{(1)}| \le \frac{(x/\alpha)\,|x/\alpha - 1|}{\Gamma(x/\alpha)\,\Gamma(2 - x/\alpha)} \cdot \frac{(1 + o(1))}{(n/\alpha)^{1+x/\alpha}} \sum_{j=(n/\alpha)^\nu}^{n/\alpha} \frac{1}{j^{1-3x/\alpha}},$$

$$|S_2^{(2)}| \le \frac{(x/\alpha)\,|x/\alpha - 1|}{\Gamma(x/\alpha)\,\Gamma(2 - x/\alpha)} \cdot \frac{(1 + o(1))}{(n/\alpha)^{1-x/\alpha}} \sum_{j=(n/\alpha)^\nu}^{n/\alpha} \frac{1}{j^{1+x/\alpha}}.$$

These inequalities imply the asymptotic estimates

$$(3.10) \qquad S_2^{(1)} = O\left(S_3\, n^{-x/\alpha} \right),$$

$$(3.11) \qquad S_2^{(2)} = O\left(S_3\, n^{-x\nu/\alpha} \right),$$

which are uniform for x from the given domain. Now (3.3) follows from equalities (3.6) and (3.9) and estimates (3.7)–(3.8) and (3.10)–(3.11). $\qquad\square$

Putting $x = 1$ in (3.2) and (3.3) and taking (0.23) into account, we arrive at the following statement.

Corollary 1 *As $n \to \infty$,*

$$C_n(\alpha,\,\alpha) = \frac{n!}{\Gamma(1/\alpha)\,(n/\alpha)^{1-1/\alpha}}\left(1 + o(1)\right), \qquad \alpha \mid n\,,$$

$$C_n(\alpha,\,\alpha/2) = \frac{n!\ 2^{2/\alpha-1}}{\Gamma(1/\alpha)\,(n/\alpha)^{1-1/\alpha}}\left(1 + o(1)\right), \qquad \frac{\alpha}{2} \mid n\,.$$

From the lemma, Corollary 1 and formula (3.1) we deduce one more corollary.

Corollary 2 *The generating functions of the random variables $\xi_n(\alpha,\,\alpha)$ and $\xi_n(\alpha,\,\alpha/2)$ have the following asymptotic representations as $n \to \infty$:*

$$(3.12) \qquad P_n(x;\alpha,\,\alpha) = \frac{\Gamma(1/\alpha)}{\Gamma(x/\alpha)} \cdot \left(\frac{n}{\alpha}\right)^{(x-1)/\alpha}\left(1 + o(1)\right), \qquad \alpha \mid n\,,$$

$$(3.13) \qquad P_n(x;\alpha,\,\alpha/2) = \frac{\Gamma(1/\alpha)}{\Gamma(x/\alpha)} \cdot \left(\frac{4n}{\alpha}\right)^{(x-1)/\alpha}\left(1 + o(1)\right), \qquad \frac{\alpha}{2} \mid n\,,$$

where $o(1) \to 0$ uniformly for $x \in [1 - \delta,\, 1 + \delta]$, $0 < \delta < 1$.

Applying Corollary 2 we prove the following theorem.

Theorem 3.1 *Let $\alpha = \alpha(n)$ be a function taking positive integer values such that $\frac{\alpha}{2}$ divides n:*

(a) If

$$\frac{1}{\alpha}\,\log\frac{n}{\alpha} \longrightarrow \lambda < \infty \qquad as \qquad n \to \infty,$$

then the limiting distribution of $\xi_n(\alpha,\,\frac{\alpha}{2})$ is a Poisson distribution with parameter λ.

(b) *If*

$$\frac{1}{\alpha} \log \frac{n}{\alpha} \longrightarrow \infty \qquad as \qquad n \to \infty,$$

then the limiting distribution of the random variable

$$\eta_n(\alpha,\, \alpha/2) = \left(\xi_n(\alpha,\, \alpha/2) - \frac{1}{\alpha} \log \frac{n}{\alpha} \right) \left(\frac{1}{\alpha} \log \frac{n}{\alpha} \right)^{-1/2}$$

is the standard normal distribution.

Similar statements are valid for the random variable $\xi_n(\alpha, \alpha)$ provided that the function $\alpha = \alpha(n)$ takes positive integer values and α divides n.

Proof We denote by $M_n(t; \alpha, \alpha/2)$ the moment generating function of the random variable $\xi_n(\alpha, \alpha)$. From (3.13), as $n \to \infty$, it follows that

$$M_n(t; \alpha,\, \alpha/2) = \frac{\Gamma(1/\alpha)}{\Gamma(e^t/\alpha)} \cdot \left(\frac{4n}{\alpha} \right)^{(e^t-1)/\alpha} \left(1 + o(1)\right),$$

where $o(1) \to 0$ uniformly for $t \in [-\delta, \delta]$, $\delta > 0$. Hence we see that, given condition (a),

$$\lim_{n \to \infty} M_n(t; \alpha,\, \alpha/2) = \exp\left\{ \lambda\left(e^t - 1 \right) \right\}$$

for any fixed $t \in [-\delta, \delta]$. The right-hand side of the equality is the moment generating function of the Poisson distribution with parameter λ and statement (a) now follows from Curtiss' theorem.

From (3.13) it is not difficult to deduce the following asymptotic representation for the moment generating function of $\eta_n(\alpha, \alpha/2)$:

$$\widetilde{M}_n(t; \alpha,\, \alpha/2) = \frac{e^{-t\sigma}\, \Gamma(1/\alpha)}{\Gamma(e^{t/\alpha}\alpha^{-1})} \cdot \left(\frac{4n}{\alpha} \right)^{(e^{t/\sigma}-1)/\alpha} \left(1 + o(1)\right),$$

where $\sigma^2 = \frac{1}{\alpha} \log \frac{n}{\alpha}$ and $o(1) \to 0$, as $n \to \infty$, uniformly for $t \in [-\delta', \delta']$, $\delta' > 0$.

This representation implies that

$$\lim_{n \to \infty} \widetilde{M}_n(t; \alpha,\, \alpha/2) = e^{t^2/2}$$

for any fixed $t \in [-\delta', \delta']$. Since $e^{t^2/2}$ is the moment generating function of the standard normal distribution, statement (b) of the theorem follows from Curtiss' theorem. $\square$

To prove the corresponding statements for the random variable $\xi_n(\alpha, \alpha)$, one can proceed in a similar way by using formula (3.12).

5.4 Extreme points in the space of symmetric stochastic matrices

A square matrix $A = \| a_{ij} \|$, $i, j = 1, 2, \ldots, n$, is said to be *stochastic* if $a_{ij} \geq 0$ and $\sum_{j=1}^{n} a_{ij} = 1$, $i = 1, 2, \ldots, n$. A stochastic matrix A is said to be *double stochastic* if $\sum_{i=1}^{n} a_{ij} = 1$, $j = 1, 2, \ldots, n$. The set Ω_n of the n by n double stochastic matrices is a convex $(n-1)^2$-dimensional space, since, for any $A, B \in \Omega_n$ and $0 \leq \alpha \leq 1$, we have $\alpha A + (1 - \alpha) B \in \Omega_n$. According to Birkhoff's theorem any matrix $A \in \Omega_n$ admits a representation (see [104])

$$(4.1) \qquad A = \alpha_1 \Pi_1 + \cdots + \alpha_s \Pi_s, \qquad \alpha_1 + \cdots + \alpha_s = 1,$$

where $\Pi_1, \ldots, \Pi_s$ are permutation matrices and $\alpha_1 \geq 0, \ldots, \alpha_s \geq 0$.

The set of permutation matrices constitutes the set of extreme points of the space Ω_n, and there is no permutation matrix A having a representation of the form (4.1) by other permutation matrices.

We consider the set Δ_n of the symmetric stochastic matrices $A = \| a_{ij} \|$, that is, such that $a_{ij} = a_{ji}$ for $i, j = 1, \ldots, n$. It is clear that Δ_n is a subset of Ω_n. We denote by Θ_n the set of extreme points of Δ_n and give a description of the matrices belonging to Θ_n.

With each permutation s of order n one can associate a matrix $C(s)$ from Θ_n in the following way. Let $s = s_1 s_2 \ldots s_k$ be the resolution of s into disjoint cycles. If (j) is a cycle of length 1, we put 1 in position (j, j) of the matrix $C(s)$; if (i, j) is a cycle of length 2, we put 1s in positions (i, j) and (j, i). If $(j_1, j_2, j_3, j_4, \ldots, j_{2l-1}, j_{2l})$ is a cycle of even length, we put 1s either in positions (j_1, j_2), (j_2, j_1), (j_3, j_4), (j_4, j_3), $\ldots, (j_{2l-1}, j_{2l})$, (j_{2l}, j_{2l-1}) or in positions (j_2, j_3), (j_3, j_2), $\ldots, (j_{2l}, j_1)$, (j_1, j_{2l}) (hence in this case s does not define $C(s)$ in a unique way). Finally, if $(j_1, j_2, j_3, j_4, \ldots, j_{2l}, j_{2l+1})$ is of odd length, we put $\frac{1}{2}$ in positions (j_1, j_2), (j_2, j_3), $\ldots, (j_{2l}, j_{2l+1})$, (j_{2l+1}, j_1) and (j_2, j_1), (j_3, j_2), $\ldots, (j_{2l+1}, j_{2l})$, (j_1, j_{2l+1}).

All the remaining elements of $C(s)$ are taken to be 0. One can show that

$$\Theta_n = \big\{ C(s) \colon s \in \mathbb{S}_n \big\}$$

where $\mathbb{S}_n$ is the symmetric group of order n [59]. By each matrix $C(s) \in \Theta_n$ we can find a unique set of $2^{\alpha_3 + \alpha_5 + \cdots}$ permutations belonging to a certain cyclic class $\big\{ 1^{\alpha_1} 2^{\alpha_2} 3^{\alpha_3} 5^{\alpha_5} \cdots \big\}$, with cycles of length 2 as the only cycles of even length. Indeed, that part of the matrix $C(s)$ corresponding to a cycle s_i of even length $l > 2$ is arranged in the same way as in the case where s_i is replaced by $l/2$ cycles of length 2 constructed by a sequential partitioning of the elements of s_i into disjoint pairs. In addition, we fix one of the possible ways of putting units into $C(s)$ for s_i and the other for s_i^{-1}. If s_i is a cycle of odd length then one and the same set of elements in $C(s)$, each of which is equal to $\frac{1}{2}$, corresponds to s_i and s_i^{-1}.

The arguments above give the following formula for the number of elements in Θ_n:

(4.2) $|\Theta_n|$

$$= \sum_{1\,\alpha_1+2\,\alpha_2+3\,\alpha_3+5\,\alpha_5+\cdots=n} \frac{n!\,2^{-(\alpha_3+\alpha_5+\cdots)}}{1^{\alpha_1}\,2^{\alpha_2}\,3^{\alpha_3}\,5^{\alpha_5}\,\cdots\,\alpha_1!\,\alpha_2!\,\alpha_3!\,\alpha_5!\,\cdots}.$$

Using the cyclic indicator $C_n(x_1,\dots,x_n)$ we can write

$$|\Theta_n| = n!\,D_n = C_n(1,\,1,\,\tfrac{1}{2},\,0,\,\tfrac{1}{2},\,0,\,\dots).$$

Therefore,

$$\sum_{n=0}^{\infty} D_n\,t^n = \exp\left\{ t + \frac{t^2}{2} + \frac{1}{2}\sum_{k=1}^{\infty} \frac{t^{2k+1}}{2k+1} \right\}.$$

Hence it follows that

$$\sum_{n=0}^{\infty} D_n\,t^n = \left(\frac{1+t}{1-t}\right)^{1/4} e^{(t+t^2)/2}, \qquad |t| < 1.$$

We consider the generating functions

(4.3) $$e^{(t+t^2)/2} = \sum_{n=0}^{\infty} u_n\,t^n,$$

(4.4) $$\left(\frac{1+t}{1-t}\right)^{1/r} = \sum_{n=0}^{\infty} v_n^{(r)}\,t^n, \qquad |t| < 1,$$

and put

(4.5) $$\sum_{n=0}^{\infty} D_n^{(r)}\,t^n = \left(\frac{1+t}{1-t}\right)^{1/r} e^{(t+t^2)/2}.$$

Obviously,

(4.6) $$D_n^{(r)} = \sum_{k=0}^{n} u_{n-k}\,v_k^{(r)}, \qquad n = 0,\,1,\,\dots;$$

and the relations

$$u_n = \sum_{j=1}^{[n/2]} \frac{1}{2^{n-j}\,j!\,(n-2j)!}$$

and

$$(4.7) \qquad v_n^{(r)} = \sum_{j=1}^{n} \binom{1/r}{n-j}\binom{1/r+j-1}{j}$$

are valid. Setting $r = 4$ we find that

$$(4.8) \quad |\Theta_n| = \sum_{k=0}^{n} \sum_{j=0}^{[(n-k)/2]} \sum_{l=0}^{k} \binom{1/4}{k-l}$$

$$\times \binom{1/4+l-1}{l} \frac{n!}{2^{n-k-j}\, j!\, (n-k-2j)!}\,.$$

Formula (4.8), although simpler than (4.2), is still awkward for large n. And so, it is natural to seek an asymptotic formula for $|\Theta_n|$ as $n \to \infty$. The following theorem is valid.

Theorem 4.1   *The number $|\Theta_n|$ of extreme points of the space Δ_n can be calculated, as $n \to \infty$, by the formula*

$$(4.9) \qquad |\Theta_n| = \frac{2^{1/4}\, e\, n!}{\Gamma(1/4)\, n^{3/4}} \left(1+o(1)\right).$$

We first prove several auxiliary lemmas.

Lemma 1   *The quantities u_k defined by (4.3) have the following asymptotic representation:*

$$(4.10) \quad u_k = \frac{1}{2\sqrt{\pi}\, k^{(k+1)/2}} \exp\left\{\frac{k+\sqrt{k}}{2}-\frac{1}{16}\right\}\left(1+o(1)\right), \quad k \to \infty.$$

Proof   Using the explicit expression (4.3) for the generating function of the sequence $\{u_k\}$ and applying Corollary to Theorem 4.1 of Chapter 1, we find, as $k \to \infty$, that

$$(4.11) \qquad u_k = \exp\left\{\tfrac{1}{2}(r^2+r)\right\} r^{-k} (4\pi r^2 + \pi r)^{-1/2} \left(1+o(1)\right),$$

where r is the unique (for large k) positive root of the equation

$$(4.12) \qquad r^2 + \tfrac{1}{2}r = k\,.$$

From (4.12) it follows, as $k \to \infty$, that

$$(4.13) \qquad r = \sqrt{k} - \frac{1}{4} + \frac{1}{32\sqrt{k}} + O\!\left(\frac{1}{k}\right),$$

$$(4.14) \qquad r^k = k^{k/2}\, e^{-\sqrt{k}/4}\left(1+o(1)\right),$$

$$\frac{1}{2}(r^2+r) = \frac{1}{2}k + \frac{1}{4}\sqrt{k} - \frac{1}{16} + o(1)\,.$$

Substituting these asymptotic expansions into formula (4.11) we see that Lemma 1 is valid.   □

Let $w_n(x)$, $n = 1, 2, \ldots$, be the sequence of functions specified by the generating function

$$(4.15) \qquad \sum_{n=0}^{\infty} w_n(x) \, t^n = \exp\left\{ (x - \tfrac{1}{2}) t + \tfrac{1}{2} x^2 t^2 \right\}.$$

Lemma 2 *If $h \to \infty$ then, uniformly for $x \in [0, 1]$,*

$$(4.16) \qquad \sum_{k=[\log n]+1}^{n} v_{n-k}^{(r)} \, w_k(x) = O\left(n^{3 - \log\log n/4} \, \log n \right),$$

where the $v_n^{(r)}$ are the same as in (4.4).

Proof From (4.3) and (4.15) it follows that

$$w_k(x) = \sum_{j=0}^{k} \frac{u_j}{2^{k-j} \, (k-j)!} \, x^j \, (x-1)^{k-j}.$$

Hence we have

$$|w_k(x)| \leq \sum_{j=0}^{[k/2]} \frac{1}{2^{k-j} \, (k-j)!} + \sum_{j=[k/2]+1}^{n} u_j$$

for all $[\log n] + 1 \leq k \leq n$ and $0 \leq x \leq 1$. Recalling Lemma 1 and the preceding estimate we calculate that, as $n \to \infty$,

$$(4.17) \qquad |w_k(x)| = O\left(n^{3 - \log\log n/4} \right)$$

uniformly for $x \in [0, 1]$ and $[\log n] + 1 \leq k \leq n$.

Applying estimates (3.4) and (3.5) we deduce from (4.7) that

$$(4.18) \qquad v_n^{(r)} = O\left(\log n \right).$$

Now formula (4.16) follows by the obvious combination of estimates (4.17) and (4.18).   □

Lemma 3 *If the $v_n^{(r)}$, $n = 1, 2, \ldots$, are given by (4.7), $r \geq 2$, and $k = o(n)$ as $n \to \infty$, then*

$$(4.19) \qquad v_{n-k}^{(r)} = \frac{2^{1/r}}{\Gamma(1/r) \, n^{1-1/r}} \left(1 + o(1) \right), \qquad n \to \infty.$$

Proof Combining (4.7) and (0.23) and setting $\alpha = 2$ and $x = 2/r$ in (3.3), we arrive at (4.19).   □

Lemma 4 *If $r \geq 2$ and $n \to \infty$ then, uniformly for $x \in [0,1]$,*

$$(4.20) \quad \sum_{k=0}^{[\log n]} v_{n-k}^{(r)} \, w_k(x) = \frac{2^{1/r}}{\Gamma(1/r) \, n^{1-1/r}} \, \exp\left\{ \tfrac{1}{2} x^2 + x - \tfrac{1}{2} \right\} \left(1 + o(1) \right).$$

Proof From Lemma 3 we know that

$$(4.21) \quad \sum_{k=0}^{[\log n]} v_{n-k}^{(r)} \, w_k(x) = \frac{2^{1/r}}{\Gamma(1/r) \, n^{1-1/r}} \, \sum_{k=0}^{[\log n]} w_k(x) \left(1 + o(1) \right).$$

Moreover,

$$(4.22) \quad \sum_{k=[\log n]+1}^{\infty} w_k(x) = O\left(n^{-\log \log n/5} \right)$$

and the estimates are uniform for $x \in [0, 1]$. Now, combining asymptotic estimates (4.21)–(4.22) and using (4.15) with $t = 1$, we obtain (4.20).   □

We define the functions $D_n^{(r)}(x)$ by the equality

$$\sum_{n=0}^{\infty} D_n^{(r)}(x) \, t^n = \left(\frac{1+t}{1-t} \right)^{1/r} \exp\left\{ \left(x - \tfrac{1}{2} \right) t + \tfrac{1}{2} x^2 t^2 \right\}.$$

Lemma 5 *If $r \geq 2$ and $n \to \infty$ then, uniformly for $x \in [0, 1]$, the asymptotic representation*

$$D_n^{(r)}(x) = \frac{2^{1/r}}{\Gamma(1/r) \, n^{1-1/r}} \, \exp\left\{ \tfrac{1}{2} x^2 + x - \tfrac{1}{2} \right\} \left(1 + o(1) \right)$$

is valid.

Proof From (4.4) and (4.15) it follows that

$$D_n^{(r)}(x) = \sum_{k=0}^{n} v_{n-k}^{(r)} \, w_k(x) \, .$$

By applying Lemmas 2 and 4 to the right-hand side of the equality we derive the desired representation.   □

Lemma 5 with $x = 1$ and $r = 4$ yields Theorem 4.1.

We consider the uniform probability distribution on the set Θ_n and study the random variable $x_n(s)$, the number of positive elements in a random matrix $C(s) \in \Theta_n$ corresponding to a permutation s of order n. The quantity $\varkappa_n(s)$, known as the *complexity* of the matrix $C(s)$, is related to the random variable $\eta_n(s)$, the total number of elements in the cycles having length 1 or 2, by the equality

$$\varkappa_n(s) = 2n - \eta_n(s) \, .$$

We now pass to finding the exact and asymptotic distributions of the random variable $\eta_n = \eta_n(s)$.

Theorem 4.2 *The disribution of η_n converges, as $n \to \infty$, to the distribution of the sum $\eta^{(1)} + 2\eta^{(2)}$ of independent Poisson random variables $\eta^{(1)}$ and $\eta^{(2)}$ with parameters 1 and $\frac{1}{2}$ respectively. In other words, we have*

$$\lim_{n\to\infty} \mathbf{P}\{\eta_n = k\} = e^{-3/2} \sum_{j=0}^{[k/2]} \frac{1}{2^j \, j! \, (k-2j)!} .$$

Proof The generating function of η_n takes the form

$$g_n(x) = \sum_{k=0}^{n} \mathbf{P}\{\eta_n = k\} \, x^k = \frac{D_n^{(4)}(x)}{D_n^{(4)}(1)} ,$$

where

$$D_n^{(4)}(x) = \frac{1}{n!} \, C_n\left(x, \, x^2, \, \tfrac{1}{2}, \, 0, \, \tfrac{1}{2}, \, 0, \, \dots\right) .$$

Applying Lemma 5 we obtain, for $0 \leq x \leq 1$,

$$g(x) = \lim_{n\to\infty} g_n(x) = \exp\left\{ \tfrac{1}{2}\, (x-1)\, (x+3) \right\} .$$

Now the theorem follows from the fact that the generating functions of $\eta^{(1)}$ and $2\eta^{(2)}$ are e^{x-1} and $e^{(x^2-1)/2}$ respectively.

We note that if $\widetilde{\eta}_n$ is the total number of elements in the cycles of length 1 or 2 of a random equiprobable permutation then the limiting distributions of $\widetilde{\eta}_n$ and η_n coincide. Put $P_k := \lim_{n\to\infty} \mathbf{P}\{\eta_n = k\}$. From Theorem 4.2 it follows that

$$P_0 = P_1 = P_2 = \tfrac{3}{2} P_3 = \tfrac{12}{5} P_4 = \tfrac{60}{13} P_5 = e^{-3/2}.$$

Hence we deduce that the number of extreme points of Θ_n having maximal complexity is asymptotically equal to $e^{-3/2} | \Theta_n |$ as $n \to \infty$, and the number of points of complexity no less than $2n - 5$ is asymptotically equal to $0.9 \cdot | \Theta_n |$.

Let $E_n \cdot n!$ be the number of such permutations of order n whose cycles have either odd length or length 2. The ratio $\gamma_n = E_n / D_n$ may be viewed as a certain averaged characteristic of the mapping multiplicity of the set of permutations with cycles of odd length or length 2 onto the set of matrices Θ_n. Since

$$E_n = \frac{1}{n!} \, C_n(1, \, 1, \, 1, \, 0, \, 1, \, 0, \, \dots) ,$$

we have

$$\sum_{n=0}^{\infty} E_n \, t^n = \left(\frac{1+t}{1-t} \right)^{1/2} e^{t^2/2}.$$

Hence, using equality (4.4), we get

$$E_n = \sum_{j=0}^{[n/2]} \frac{v_{n-2j}^{(2)}}{2^j \, j!} \, , \qquad n = 0, 1, \ldots .$$

Applying Lemma 3 we see that

$$\sum_{j=1}^{[\log n]} \frac{v_{n-2j}^{(2)}}{2^j \, j!} = \frac{1}{\Gamma(1/2)} \, \sqrt{\frac{2\,e}{n}} \left(1 + o(1) \right) .$$

Furthermore,

$$\sum_{j=[\log n]+1}^{[n/2]} \frac{v_{n-2j}^{(2)}}{2^j \, j!} = O\left(n^{1-\log\log n} \, \log n \right) .$$

From the two preceding estimates we deduce that

$$E_n = \frac{1}{\Gamma(1/2)} \, \sqrt{\frac{2\,e}{n}} \left(1 + o(1) \right) .$$

Hence, recalling (4.9), we finally obtain

$$\gamma_n = \frac{\Gamma(1/4)}{\Gamma(1/2)} \left(\frac{2\,n}{e^2} \right)^{1/4} \left(1 + o(1) \right) . \qquad\qquad \Box$$

CHAPTER 6

Random Graphs and Random Mappings

6.0 Introduction

This chapter is devoted to random graphs. A number of ways of introducing randomness for various classes of graph exists, one of which is to specify on some classes of graph (trees, forests, graphs of one-to-one mappings and so on) certain, as a rule uniform, probability distributions. The second way of constructing random graphs is defined by a stochastic process which gives a rule for joining a number of initially isolated vertices by edges. The third way, which is closely related to the second, is described by a random procedure of deletion of edges from a complete graph. Other methods for constructing random graphs exist but they are of little use.

Before proceeding to describe results in the field, we list a number of statements concerning the combinatorial properties of graphs that will be required in the sequel. In this chapter we deal mainly with labeled graphs and for this reason the results cited below are related, as a rule, to such combinatorial structures.

6.0.1 Trees

Labeled trees are, in a sense, the simplest labeled graphs. A *tree* is a connected graph with no cycles. A *rooted* tree is a tree which has a distinguished vertex called the *root*. A labeled tree which is not a rooted tree is called an *unrooted labeled* tree. If r_n is the number of labeled rooted trees with n vertices and $\tilde{r}_n$ is the number of unrooted labeled trees with n vertices then for $n = 2, 3, \ldots$ (see [111])

$$(0.1) \qquad\qquad r_n = n^{n-1}, \qquad\qquad \tilde{r}_n = n^{n-2}.$$

A graph consisting of several trees is called a *forest*.

If $\tilde{r}_{nk}$ is the number of unrooted labeled forests, each of which has $n+k-1$ vertices and k trees and where k fixed vertices belong to distinct unrooted labeled trees, then the following formula is valid [18]:

$$(0.2) \qquad \tilde{r}_{nk} = k\,(n+k-1)^{n-2}$$

and $\tilde{r}_n = \tilde{r}_{n1}$.

If $C(n,k)$ is the number of unrooted labeled trees in which a given vertex is incident to k edges then, in view of the obvious equality

$$C(n,k) = \binom{n-1}{k}\,\tilde{r}_{n-k,\,k}\,,$$

we have

$$(0.3) \qquad C(n,k) = \binom{n-2}{k-1}\,(n-1)^{n-k-1}.$$

The number of forests consisting of k unrooted labeled trees is

$$\tilde{r}_n^{(k)} = \frac{1}{k!}\sum_{j=0}^{k}\left(-\frac{1}{2}\right)^{j}\binom{k}{j}\,(k+j)\,n^{n-k-j-1}\,(n)_{k+j}\,, \qquad k = 1, 2, \ldots, n.$$

It is known that (see [91])

$$(0.4) \qquad \lim_{n\to\infty}\frac{\tilde{r}_n^{(k)}}{n^{n-2}} = \frac{1}{2^{k-1}\,(k-1)!}\,, \qquad k = 1, 2, \ldots .$$

The number of forests consisting of k rooted labeled trees is calculated by the formula

$$(0.5) \qquad r_n^{(k)} = \binom{n-1}{k-1}\,n^{n-k}\,, \qquad k = 1, 2, \ldots, n.$$

Prufer [89] describes a method which gives a one-to-one correspondence between the set of unrooted labeled trees with n vertices and the set of all one-to-one mappings $\phi:\ :\mathbf{X}\mapsto\mathbf{Y},\ |\mathbf{X}| = n,\ |\mathbf{Y}| = n-2$, where $|\mathbf{A}|$ is the number of elements in the set $\mathbf{A}$. This correspondence is constructed by a sequential deletion of the edges of a labeled unrooted tree which are incident to the terminal vertex whose label has (at the moment) minimal value. According to this procedure $\phi(k)$ is the label of the vertex with which the edge removed at the kth step joins the terminal vertex with minimal label. The process terminates at the moment when all but one of the edges are removed and, therefore, k takes the values $1, 2, \ldots, n-2$. The number of vertices $\chi(i)$ incident with the vertex i is called the *valency* of the vertex and the number $\mu(i) := \chi(i) - 1$ is called the *degree* of the vertex. It follows from Prufer's construction that the degrees of vertices are determined by the primary specification of the corresponding mapping ϕ. The elements of the set $\{1, 2, \ldots, n\}$ not belonging to the set of images of ϕ are the labels of the terminal vertices of the tree. The correspondence reduces the problem of enumeration trees with specified tuples of degrees of vertices to that of enumeration mappings with a given primary specification or, in other words, to the allocation problem of unlike particles into unlike cells with a given specification of allocations of the cells. These arguments show, in particular, that the number of unrooted labeled trees with k terminal vertices, denoted by $\tilde{r}(n, k)$, can be calculated by the formula

$$(0.6) \qquad \tilde{r}(n, k) = \frac{n!}{k!}\, \sigma(n - 2, n - k)\,, \qquad\qquad k = 2, 3, \ldots, n - 1\,,$$

where $\sigma(n, k)$ are Stirling's numbers of the second kind defined by the relation

$$(0.7) \qquad\qquad x^n = \sum_{k=0}^{n} \sigma(n, k)\,(x)_k\,.$$

For any two vertices of a tree there exists a unique path connecting the vertices. The number of edges in the path is called the path's *length*. The *height* of a rooted tree is the length of the longest path in the tree starting from its root.

Let $D_n^{(h)}$ be the number of rooted labeled trees with n vertices whose heights do not exceed h. As shown in [95], the generating function

$$(0.8) \qquad\qquad a_h(t) = \sum_{n=1}^{\infty} D_n^{(h)}\, \frac{t^n}{n!}\,, \qquad\qquad h = 0, 1, \ldots\,,$$

satisfies the recurrence formula

$$(0.9) \qquad\qquad a_{h+1}(t) = t\, \exp\left\{a_h(t)\right\}\,, \qquad\qquad h = 0, 1, \ldots\,,$$

with $a_0(t) = t$. It follows from (0.9) that

$$\lim_{h \to \infty} a_h(t) = \sum_{n=1}^{\infty} n^{n-1}\, \frac{t^n}{n!}$$

and the series on the right-hand side of the equality is convergent for $|t| \le 1/e$ and represents the inverse function $y = y(t)$ of the functional relationship

$$(0.10) \qquad\qquad t = y\, e^{-y}\,.$$

Let a uniform distribution be given on the set of all unrooted labeled trees with n vertices, and let θ_n be the height of a randomly chosen tree with respect to a fixed vertex. Then (see [93])

$$\lim_{n \to \infty} \mathbf{P}\left\{\theta_n < x\sqrt{2n}\right\} = \sum_{v=-\infty}^{\infty} e^{-v^2 x^2}\left(1 - 2v^2 x^2\right) \equiv \mathbf{P}\left\{\theta < x\right\}.$$

Moments of the random variable θ are calculated by the formulae

$$\mathbf{E}\,\theta = \sqrt{\pi}\,,$$

$$\mathbf{E}\,\theta^s = 2\Gamma\left(\tfrac{1}{2}s + 1\right)(s-1)\,\zeta(s), \qquad s > 1,$$

where $\zeta(s) := \sum_{m=1}^{\infty} m^{-s}$. The variance of θ is

$$\mathrm{Var}\,\theta = \tfrac{1}{3}\pi\,(\pi - 3).$$

6.0.2   Graphs of single-valued mappings

Let $\mathfrak{S}_n$ be the symmetric semigroup of single-valued mappings of an n-element set $\mathbf{X}$ onto itself. The *graph of a mapping* σ is an oriented graph $\Gamma(\mathbf{X}, \sigma)$ whose vertices x and x' are joined by the edge (x, x') if $x' = \sigma(x)$. Each graph $\Gamma(\mathbf{X}, \sigma)$, $\sigma \in \mathfrak{S}_n$, consists of connected components, and each component contains one cycle and trees whose roots are the vertices of the cycle called *cyclic elements*. It is assumed that the edges of the random trees are directed to the roots. Let $\mathbf{A}$ be a subset of the set $\mathbb{N} = \{1, 2, \dots\}$ of all positive integers and let $\mathfrak{S}_n(\mathbf{A})$ denote the collection of mappings $\sigma \in \mathfrak{S}_n$ for which the lengths of the corresponding cycles are elements of $\mathbf{A}$. The elements of the set $\mathfrak{S}_n(\mathbf{A})$ are called $\mathbf{A}$-*mappings*. Let $U_n(k; \mathbf{A})$ and $U_{nj}(\mathbf{A})$ represent the number of $\mathbf{A}$-mappings with k cyclic elements and with j components respectively. Then

$$(0.11) \quad U_n(k; \mathbf{A}) = \binom{n-1}{k-1} n^{n-k}\, C(k; \mathbf{A}), \qquad\qquad k = 0, 1, \dots,$$

$$(0.12) \quad U_{nj}(\mathbf{A}) = \sum_{k=j}^{n} \binom{n-1}{k-1} n^{n-k}\, C(k, j; \mathbf{A}), \qquad j = 0, 1, \dots,$$

where $U_0(0; \mathbf{A}) := U_{00}(\mathbf{A}) = 1$ and $C(k; \mathbf{A})$ and $C(k, j; \mathbf{A})$ are the numbers of $\mathbf{A}$-permutations and $\mathbf{A}$-permutations with j cycles of degree k respectively. In particular, setting

$$U_n(k) := U_n(k; \mathbb{N}), \qquad\qquad U_{nj} := U_{nj}(\mathbb{N}),$$

we have

(0.13) $U_n(k) = k\,n^{n-k}\,(n-1)_{k-1}\,,$ $k = 1, 2, \dots ,$

(0.14) $U_{nj} = \displaystyle\sum_{k=j}^{n} \binom{n-1}{k-1} n^{n-k}\,|\,s(k,j)\,|\,,$ $j = 1, 2, \dots ,$

where $s(k,j)$ are Stirling's numbers of the first kind. Let $T_n^{(h)}(k,j;\mathbf{A})$ denote the number of $\mathbf{A}$-mappings with k cyclic elements and j components such that the height of any tree of the corresponding graphs does not exceed h. Then [111]

$$
(0.15) \qquad \exp\left\{z\,A\big(x\,a_h(t)\big)\right\} = \sum_{n=0}^{\infty} \sum_{k=0}^{n} \sum_{j=0}^{k} T_n^{(k)}(k,j;\mathbf{A})\,\frac{t^n}{n!}\,x^k\,z^j ,
$$

where

$$
A(w) := \sum_{j \in \mathbf{A}} \frac{w^j}{j}
$$

and $a_h(t)$ is the generating function defined by (0.8). The generating functions for the sequences $T_n^{(h)}(k;\mathbf{A})$ and $T_{nj}^{(h)}(\mathbf{A})$ of the numbers of the corresponding $\mathbf{A}$-mappings with k cyclic elements and with j components can be obtained from (0.15) by setting $z = 1$ and $x = 1$ respectively.

6.1 Random trees and forests

We consider the uniform distributions on the sets of forests consisting of rooted labeled trees and of unrooted labeled trees, respectively, and study various characteristics of the random objects obtained. In this section we find the exact and limiting distributions of such characteristics.

6.1.1 The number of trees in a random forest

The limiting distribution of the number of trees in a random forest is described by the following theorem.

Let ξ_n and $\tilde{\xi}_n$ be the numbers of rooted and unrooted trees, respectively, in a random labeled forest with n vertices. Put $\eta_n := \xi_n - 1$ and $\tilde{\eta}_n := \tilde{\xi}_n - 1$.

Theorem 1.1 *As $n \to \infty$, the distribution of the random variable $\eta_n\,(\tilde{\eta}_n)$ converges to a Poisson distribution with parameter $\lambda = 1\,(\tilde{\lambda} = 1/2)$.*

Proof It follows from (0.5) that the number of rooted labeled forests with n vertices is equal to $(n+1)^{n-1}$ and thus the exact distribution of ξ_n is

$$(1.1) \qquad \mathbf{P}\{\xi_n = k\} = \binom{n-1}{k-1} \frac{n^{n-k}}{(n+1)^{n-1}}, \qquad k = 1, 2, \ldots, n.$$

This equality gives the following expression for the generating function of η_n:

$$\varphi_n(x) = \left(1 + \frac{x-1}{n+1}\right)^{n-1}.$$

Hence we see that, for any fixed x,

$$\lim_{n\to\infty} \varphi_n(x) = e^{x-1}.$$

This proves the desired statement for η_n.

To prove the corresponding statement for $\tilde{\eta}_n$ we observe that, according to (0.4), the number of unrooted labeled forests with n vertices is asymptotically equal to $e^{-1/2} n^{n-2}$. Therefore,

$$(1.2) \qquad \lim_{n\to\infty} \mathbf{P}\{\tilde{\eta}_n = k\} = \frac{1}{2^k\, k!}\, e^{-1/2}, \qquad k = 0, 1, \ldots.$$

This completes the proof of the theorem. $\qquad\qquad\qquad\qquad\qquad\square$

6.1.2 Degrees of vertices of a tree

We recall that the degree of a vertex is equal to the number of vertices incident to it decreased by one. If μ_n is the degree of a vertex of an unrooted labeled tree then, according to (0.3), we have

$$(1.3) \qquad \mathbf{P}\{\mu_n = k\} = \binom{n-2}{k} \frac{(n-1)^{n-k-2}}{n^{n-2}}, \qquad k = 0, 1, \ldots, n-2.$$

The generating function of μ_n takes the form

$$(1.4) \qquad P(x; \mu_n) = \left(1 - \frac{1}{n}\right)^{n-2} \left(1 + \frac{x}{n-1}\right)^{n-2}.$$

Hence it is easy to deduce the following theorem.

Theorem 1.2 *The distribution of the degree μ_n of a fixed vertex of a random unrooted unlabeled tree converges to a Poisson distribution with parameter $\lambda = 1$ as $n \to \infty$.*

We denote by $\nu_n^{(s)}$ the number of vertices of degree s in a random unlabeled tree. In view of Prufer's correspondence described in the introduction to this chapter the distribution of $\nu_n^{(s)}$ coincides with that of the number of such elements of a noncommutative, nonsymmetric n-basis occurring exactly s times in a random $(n-2)$-sample, which means that, in fact, the limiting distributions of $\nu_n^{(s)}$, as $n \to \infty$, have already been studied in Chapter 3. According to Theorem 3.3 of Chapter 3, if we put

$$\theta_1(s) = \frac{e^{-1}}{s!}\,, \qquad \theta_2(s) = \theta_1(s)\left[1 - \theta_1(s)\left(1 + (1-s)^2\right)\right],$$

then the distribution of the random variable $(\nu_n^{(s)} - \theta_1(s)\,n)\,(\theta_2(s)\,n)^{-1/2}$ is asymptotically normal with parameters $(0,1)$ as $n \to \infty$.

Here we consider only the limiting distribution of the random variable $\nu_n = \nu_n^{(0)}$, the number of terminal vertices, and apply a method from [91], which differs from those of Chapter 3, to find the distribution.

Theorem 1.3 *The distribution of the number ν_n of the terminal vertices of a random unrooted tree has asymptotically normal distribution as $n \to \infty$, namely,*

$$\lim_{n\to\infty} \mathbf{P}\left\{\frac{\nu_n - n\,e^{-1}}{\sqrt{n\,(e-2)\,e^{-1}}} < x\right\} = \frac{1}{\sqrt{2\pi}} \int_{-\infty}^{x} e^{-u^2/2}\,du\,.$$

Proof Because of (0.6) we have

$$(1.5) \quad \mathbf{P}\{\nu_n = k\} = \frac{n!}{n^{n-2}\,k!}\,\sigma(n-2, n-k), \qquad k = 2, 3, \ldots .$$

From (0.7) it follows that

$$(1.6) \qquad \sum_{k=2}^{n-1} \mathbf{P}\{\nu_n = k\}\,\frac{(x)_{n-k}}{(n)_{n-k}} = \left(\frac{x}{n}\right)^{n-2}.$$

Setting $x = n-1$ and $x = n-2$ in this equality we are able to deduce without difficulty the following formulae for the mean and variance of ν_n:

$$\mathbf{E}\,\nu_n = n\left(1 - \frac{1}{n}\right)^{n-2}\,,$$

$$\mathrm{Var}\,\nu_n = (n)_2\left(1 - \frac{2}{n}\right)^{n-2} + n\left(1 - \frac{1}{n}\right)^{n-2} - n^2\left(1 - \frac{1}{n}\right)^{2n-4}.$$

Letting $n \to \infty$ we see that

$$(1.7) \qquad\qquad \mathbf{E}\,\nu_n = \frac{n}{e}\left(1 + o(1)\right),$$

$$(1.8) \qquad\qquad \mathrm{Var}\,\nu_n = \frac{n(e-2)}{e^2}\left(1 + o(1)\right).$$

By merely setting $x = n - it\sqrt{n}$ in (1.6) we find

$$(1.9) \qquad \sum_{k=2}^{n-1} \mathbf{P}\{\nu_n = k\} \prod_{j=k+1}^{n} \left(1 - \frac{it\, n^{1/2}}{j}\right) = \left(1 - \frac{it}{\sqrt{n}}\right)^{n-2}.$$

If $n \to \infty$ and $k = o(\sqrt{n})$ then

$$(1.10) \qquad \mathbf{P}\{\nu_n = k\} = \frac{n!}{k!} \cdot \frac{(n-k)^{2(k-2)}}{2^{k-2}\,(k-2)!\,n^{n-2}} \left(1 + o(1)\right).$$

This equality can be obtained by (1.5) and the following asymptotic relation for Stirling's numbers of the second kind [57]:

$$(1.11) \qquad \sigma(n,k) = \frac{k^{2(n-k)}}{2^{n-k}\,(n-k)!} \left(1 + o\left(\frac{(n-k)^2}{n}\right)\right),$$

which is valid if $n, k \to \infty$ and $n - k = o(\sqrt{n})$. Taking into account the boundedness of the quantity

$$\left| \prod_{j=k+1}^{n} \left(1 - \frac{it\, n^{1/2}}{j}\right) \right|$$

as $n, k \to \infty$, and estimating (1.10) we conclude that, for any fixed t,

$$(1.12) \qquad \left| \sum_{k=2}^{[\log n]} \mathbf{P}\{\nu_n = k\} \prod_{j=k+1}^{n} \left(1 - \frac{it\, n^{1/2}}{j}\right) \right| < \exp\left\{-n + C_1 \log^2 n\right\},$$

where C_1 is a positive constant. Further, by applying Chebyshev's inequality for $k > [\log n]$ it is not difficult to show that

$$(1.13) \qquad \left| \sum_{\substack{|k - n\,e^{-1}| > n^{\beta} \\ k > [\log n]}} \mathbf{P}\{\nu_n = k\} \prod_{j=k+1}^{n} \left(1 - \frac{it\, n^{1/2}}{j}\right) \right| = O(n^{1-2\beta}),$$

where $\frac{1}{2} < \beta < \frac{3}{4}$. Finally, for all k satisfying $|k - n\,e^{-1}| < n^{\beta}$, we have

$$(1.14) \qquad \log \prod_{j=0}^{n-k-1} \left(1 - \frac{it\, n^{1/2}}{n-j}\right)$$

$$= -it\, n^{1/2} + it\, n^{-1/2}\,(ek - n) + \tfrac{1}{2} t^2\,(e - 1) + o(1).$$

Estimates (1.12)–(1.14) yield

(1.15) $\lim\limits_{n\to\infty} \Bigg| \left(1 - itn^{-1/2}\right)^{n-2} \exp\left\{itn^{1/2}\right\}$

$$- \exp\left\{\tfrac{1}{2}t^2(e-1)\right\} \sum_{k=2}^{n-1} \mathbf{P}\left\{\nu_n = k\right\} \exp\left\{itn^{-1/2}(ek - n)\right\} \Bigg| = 0.$$

Since

$$\lim_{n\to\infty} \left(1 - itn^{-1/2}\right)^{n-2} e^{it\sqrt{n}} = e^{t^2/2},$$

relation (1.15) implies the equality

$$\lim_{n\to\infty} \exp\left\{-it\sqrt{\frac{n}{e-2}}\right\} \sum_{k=2}^{n-1} \mathbf{P}\left\{\nu_n = k\right\} \exp\left\{\frac{itke}{\sqrt{n(e-2)}}\right\} = e^{-t^2/2}.$$

This completes the proof of the theorem.   □

6.1.3 The distance between vertices in a tree

We denote by γ_n the number of vertices of a random unrooted tree in the path connecting two fixed vertices and let d_n be the distance between the vertices. It is clear that $d_n = \gamma_n - 1$. The exact distribution of the random variable γ_n is given by the formula [73]

(1.16) $\mathbf{P}\left\{\gamma_n = k\right\} = \dfrac{k}{n-1}\,\dfrac{(n)_k}{n^k}\,,$ $k = 2, 3, \ldots, n.$

Indeed, one can choose and arrange the vertices of a path of length k connecting two fixed vertices in $\binom{n-2}{k-2}(k-2)!$ ways. The number of trees with n vertices containing a path passing through k fixed vertices is $k\,n^{n-k-1}$. Multiplying these quantities and dividing the product obtained by n^{n-2} we get formula (1.16).

Theorem 1.4 *Let* $n \to \infty$ *and* $u = k/\sqrt{n}$, $k = 1, 2, \ldots$, *vary with* n *in such a way that* $u = o(n^{1/6})$. *Then*

(1.17) $\mathbf{P}\left\{\dfrac{\gamma_n}{\sqrt{n}} = u\right\} = \dfrac{u}{\sqrt{n}}\,e^{-u^2/2}\left(1 + o(1)\right).$

Furthermore, for any fixed x,

(1.18) $\lim\limits_{n\to\infty} \mathbf{P}\left\{\dfrac{\gamma_n}{\sqrt{n}} < x\right\} = 1 - e^{-x^2/2}.$

Proof By applying Stirling's formula we find

$$\mathbf{P}\left\{\frac{\gamma_n}{\sqrt{n}} = u\right\} = \frac{u}{\sqrt{n}}\, e^{-u\sqrt{n}}\left(1 - \frac{u}{\sqrt{n}}\right)^{-n+u\sqrt{n}-1/2}\left(1 + o(1)\right).$$

For all $u = o(n^{1/6})$ the following estimate is valid:

$$\left(-n + u\, n^{1/2} - \tfrac{1}{2}\right)\log\left(1 - u\, n^{-1/2}\right) = u\, n^{1/2} - \tfrac{1}{2}\, u^2 + o(1).$$

The preceding two estimates imply equality (1.17). We now observe that the asymptotic density of the distribution of the random variable $\gamma_n/\sqrt{n}$ is $u\exp\{-u^2/2\}$. By integrating the density over the interval $[0, x]$ we obtain (1.18). $\qquad\qquad\qquad\qquad\qquad\qquad\qquad\qquad\qquad\qquad\qquad\square$

The generating function of γ_n can be represented as

$$f_n(x) = x^2 + n!\left(\frac{x - 1}{n - 1}\right)\sum_{k=0}^{n-3}\frac{1}{k!}\left(\frac{x}{n}\right)^{n-k-1}.$$

By differentiating $f_n(x)$ at the point $x = 1$ one can find the mean and variance of γ_n:

$$\mathbf{E}\,\gamma_n = 1 + L_n,$$

$$\mathrm{Var}\,\gamma_n = 2\,n - 3\,L_n - L_n^2,$$

where

(1.19) $$L_n := \frac{(n-2)!}{n^{n-2}}\sum_{k=0}^{n-2}\frac{n^k}{k!}.$$

Using the asymptotic formula

(1.20) $$e^{-n}\sum_{k=0}^{n}\frac{n^k}{k!} = \frac{1}{2} + o(1)$$

we see that

(1.21) $$\mathbf{E}\,\gamma_n = \sqrt{\frac{\pi\, n}{2}}\,\left(1 + o(1)\right),$$

(1.22) $$\mathrm{Var}\,\gamma_n = \left(2 - \frac{\pi}{2}\right) n\,\left(1 + o(1)\right).$$

It is clear that, by the equality $d_n = \gamma_n - 1$, all the asymptotic results concerning γ_n can immediately be extended to d_n.

6.1.4 The mass of a vertex of a tree

With any vertex of an unrooted tree which is labeled by a symbol i, $1 \leq i < n$, say, we associate a subtree which includes all the vertices j of the initial tree such that the paths joining the vertices labeled by n and j pass through i. The number of vertices in the subtree is called the *mass* of the vertex i. We will find the distribution of the mass μ_n of a fixed vertex in a random unrooted tree with n vertices.

There are $\binom{n-2}{k-1} k^{k-2}$ ways of choosing labels and constructing a subtree which determines the mass of a fixed vertex. Having constructed such a subtree one can complete the construction of the tree in $(n-k)^{n-k-1}$ ways. Thus we have

$$(1.23) \qquad \mathbf{P}\{\mu_n = k\} = \binom{n-2}{k-1} \frac{k^{k-2}}{n^{n-2}} (n-k)^{n-k-1},$$

$$k = 1, 2, \ldots, n-1.$$

Theorem 1.5 *If* $n \to \infty$ *and* $k = o(n)$ *then*

$$(1.24) \qquad \mathbf{P}\{\mu_n = k\} = \frac{k^{k-1}}{k!} e^{-k} \left(1 + o(1)\right).$$

To prove the theorem it suffices to apply Stirling's formula to the right-hand side of (1.23).

With the aid of the identities

$$\frac{1}{n^{n-2}} \sum_{j=0}^{n-2} \binom{n-2}{j} (j+1)^j (n-j-1)^{n-j-2} = \frac{n}{n-1} \sum_{j=2}^{n} \frac{(n)_j}{n^j},$$

$$\frac{1}{n^{n-2}} \sum_{j=0}^{n-2} \binom{n-2}{j} (j+1)^{j+1} (n-j-1)^{n-j-2} = \frac{n}{n-1} \sum_{j=2}^{n} \binom{j}{2} \frac{(n)_j}{n^j},$$

obtained by Riordan [97], and the equality

$$\sum_{j=2}^{n} \binom{j}{2} \frac{(n)_j}{n^j} = \frac{n}{2} \sum_{j=2}^{n} \frac{(n)_j}{n^j},$$

one can find explicit expressions for the first two moments of μ_n [77]:

$$(1.25) \qquad \mathbf{E}\,\mu_n = \frac{n}{n-1} \sum_{j=2}^{n} \frac{(n)_j}{n^j},$$

$$(1.26) \qquad \mathbf{E}\,\mu_n^2 = \frac{n^2}{2(n-1)} \sum_{j=2}^{n} \frac{(n)_j}{n^j}.$$

By applying the obvious identity

(1.27) $$\sum_{j=2}^{n} \frac{(n)_j}{n^j} = \frac{n!}{n^n} \sum_{j=0}^{n} \frac{n^j}{j!} - 2$$

and formula (1.20) we finally obtain, as $n \to \infty$,

(1.28) $$\mathbf{E}\,\mu_n = \sqrt{\frac{\pi n}{2}}\left(1 + o(1)\right),$$

(1.29) $$\mathbf{E}\,\mu_n^2 = \sqrt{\frac{\pi n^3}{8}}\left(1 + o(1)\right).$$

6.1.5 Cuts of trees and random walks

Let the vertices of an unrooted tree be labeled by the elements of an n-element set $\mathbf{X}$. The *cutting number* of a vertex labeled by an element $x \in \mathbf{X}$ is the number of unordered pairs of vertices (u,v), $u,v \neq x$, such that the path joining u and v passes through x. If δ_n is the cutting number of a vertex of a random unrooted tree then (see [74])

$$\mathbf{E}\,\delta_n = \frac{n}{2} \sum_{j=3}^{n} \frac{(n)_j}{n^j}.$$

Hence it follows that

(1.30) $$\mathbf{E}\,\delta_n = \sqrt{\frac{\pi n^3}{8}}\left(1 + o(1)\right), \qquad n \to \infty.$$

We now consider the set of all rooted labeled trees where the label of the root is one and the same fixed number for all the trees. For each such tree we define a *random walk* around the tree as follows. If d_x is the number of vertices incident to the vertex x then the probability of transition at the next step from x to a vertex y is

$$P_{xy} := \begin{cases} d_x^{-1}, & x \sim y, \\ 0, & \text{otherwise}, \end{cases}$$

where the relation $x \sim y$ means that the vertices labeled x and y are adjacent in the tree. The average number of steps λ_n until the first return to the root of a random rooted labeled tree is given by the formula

$$\mathbf{E}\,\lambda_n = \frac{2(n-1)^{n-1}}{n^{n-2}}\left[\left(1 + \frac{1}{n-1}\right)^{n-1} - 1\right].$$

Hence we obtain

(1.31) $$\mathbf{E}\,\lambda_n = \frac{2n(e-1)}{e}\left(1 + o(1)\right), \qquad n \to \infty.$$

6.1.6 Unlabeled trees

We now study the class of unlabeled trees. Let t_n be the number of unrooted unlabeled trees and T_n be the number of rooted unlabeled trees with n vertices. It is known that the generating functions

$$(1.32) \qquad t(x) := \sum_{n=1}^{\infty} t_n\, x^n$$

and

$$(1.33) \qquad T(x) := \sum_{n=1}^{\infty} T_n\, x^n$$

satisfy the following relations [51]:

$$(1.34) \qquad T(x) = x \exp\left\{ \sum_{j=1}^{\infty} \frac{T(x^j)}{j} \right\},$$

$$(1.35) \qquad t(x) = T(x) - \tfrac{1}{2}\, T^2(x) + \tfrac{1}{2}\, T(x^2).$$

Equality (1.34) gives the recurrence formula

$$T_n = \frac{1}{n-1} \sum_{k=1}^{n-1} T_{n-k} \sum_{m\,|\,k} m\, T_m, \qquad n > 1,$$

where $T_1 = 1$. In turn, equality (1.35) implies

$$t_n = T_n - \left[\binom{T_{n/2}}{2} + \sum_{k=1}^{\lfloor (n-1)/2 \rfloor} T_k\, T_{n-k} \right].$$

One can establish that the following asymptotic formulae are valid as $n \to \infty$ (see, for example, [85]):

$$(1.36) \qquad T_n = \frac{0.4399\ldots}{n^{3/2}\, \rho^n} \left(1 + o(1) \right),$$

$$(1.37) \qquad t_n = \frac{0.5349\ldots}{n^{3/2}\, \rho^n} \left(1 + o(1) \right),$$

where ρ is the radius of convergence of the series $T(x)$, $\rho = 0.3383\ldots$. In [99] rather complicated formulae are obtained for the distribution of degrees of vertices of random unlabeled trees which we do not give here. We only note that the probability of the event that a vertex of a random unrooted unlabeled tree with n vertices is terminal is approximately equal to $0.4381\ldots$ as $n \to \infty$.

6.1.7   Rooted recursive trees

For $n = 1$ a rooted recursive tree consists of a single vertex (root) labeled 1. A rooted recursive tree H_n with n vertices labeled $1, 2, \ldots, n$ is constructed from a recursive tree H_{n-1} with $n-1$ vertices $1, 2, \ldots, n-1$ by choosing a vertex of H_{n-1} and joining a vertex labeled n to it.

The number of recursive trees with n vertices is equal to $(n-1)!$. Obviously, a tree is recursive if and only if for each $k = 1, 2, \ldots, n$ the labels of the vertices in the unique path joining the root to the vertex labeled k form an increasing subsequence of the sequence $\{1, 2, \ldots, k\}$.

Let d_{ij} be the distance between vertices i and j in a random equiprobable recursive tree and let $P(i, j; d) := \mathbf{P}\{d_{ij} = d\}$. Then for $1 \le i < j \le n$ and $1 \le d \le n-1$ we have [78]

$$(1.38) \qquad P(i, j; d) = \frac{1}{j-1}\{P(1, i; d-1)$$

$$+ P(2, i; d-1) + \cdots + P(j-1, i; d-1)\}.$$

Indeed, if H_{j-1} is a recursive subtree determined by $j-1$ vertices and vertex j is joined by an edge to a vertex x then the distance between i and j is equal to that between i and x in the tree H_{j-1} plus 1. There exist $j-1$ equiprobable ways to select the vertex in H_{j-1} to be adjacent to j. Hence relation (1.38) follows. By the relation we find the mean value of d_{ij}:

$$\mathbf{E}\, d_{ij} = h_i + h_{j-1} + i^{-1},$$

where

$$h_1 = 0, \qquad h_k = \sum_{\nu=2}^{k} \nu^{-1}, \qquad k \ge 2.$$

The average distance B_n between two distinct vertices chosen at random is determined by the formula

$$B_n = 2\,\frac{n+1}{n-1}\,h_n - 2 = 2 \log n + O(1).$$

6.2   Random graphs

In this section we study the structure of random graphs, assuming that the number of their vertices tends to infinity. This structure is determined by a number of characteristics such as the number of connected components, the sizes of the components, the degrees of vertices and so on. These characteristics are random variables whose distributions describe the asymptotic structure of random graphs.

Let the vertices of a graph Γ be labeled by the elements of a set $\mathbf{X}$. We recall that a subgraph Γ' of the graph Γ with a set of vertices $\mathbf{X}' \subseteq \mathbf{X}$ is said to be a *connected component* of Γ if

(a) for any $x, x' \in \mathbf{X}'$ there exists a path in Γ connecting the vertices x and x';

(b) if $x'' \notin \mathbf{X}'$ then the subgraph Γ'' of Γ with the set of vertices $\mathbf{X}' \cup \{x''\}$ possesses no property (a).

A graph Γ is said to be *connected* if it has exactly one connected component.

6.2.1 Asymptotic structure of random graphs

We denote by $\Gamma_{n,N}$ a graph with n labeled vertices and N edges, $0 \leq N \leq \binom{n}{2}$. We specify the uniform distribution on the set of such graphs by assigning to each graph $\Gamma_{n,N}$ the probability $\left(\binom{n}{2} \atop N \right)^{-1}$. Put

$$(2.1) \qquad N(C) := \left[\tfrac{1}{2} n \log n + C n \right],$$

where C is a fixed real number and $[x]$ stands for the integral part of x.

We say that a graph $\Gamma_{n,N(C)}$ is *of type* $\mathfrak{A}$ if, for some k, it has a connected component with $n - k$ vertices and k isolated points. We denote by $P(\mathfrak{A}, n, N(C))$ the probability that a random graph $\Gamma_{n,N(C)}$ is of type $\mathfrak{A}$ and set

$$P(\bar{\mathfrak{A}}, n, N(C)) := 1 - P(\mathfrak{A}, n, N(C)).$$

The following lemma is due to Érdös and Rényi [30].

Basic lemma *For large* n, *almost all graphs* $\Gamma_{n,N(C)}$ *are of type* $\mathfrak{A}$, *that is,*

$$(2.2) \qquad \lim_{n \to \infty} P(\bar{\mathfrak{A}}, n, N(C)) = 0.$$

Proof We fix a number M satisfying the inequality

$$M \leq n - 2 N(C) n^{-1}$$

and partition the set of all graphs $\Gamma_{n,N(C)}$ into two classes. We attribute to the first class E_M all those graphs whose largest component contains at least $n - M$ vertices. All the remaining graphs are of the class $\bar{E}_M$. If $\Gamma_{n,N(C)} \in \bar{E}_M$ and $\Gamma_{n,N(C)}$ has r connected components of sizes $l_1, l_2, \ldots, l_r$ then $l_1 + l_2 + \cdots + l_r = n$ and $\sum_{i=1}^{r} \binom{l_i}{2} \geq N(C)$. Therefore, if $L = \max_i l_i$ then $L - 1 \geq 2 n^{-1} N(C)$ and thus $L > 2 n^{-1} N(C)$. If the largest component of a graph $\Gamma_{n,N(C)} \in \bar{E}_M$ has $n - s$ vertices then the graph has at most

$\binom{n}{2} - s(n-s)$ edges. And thus the probability that a random graph $\Gamma_{n,N(C)}$ belongs to $\bar{E}_M$ can be estimated as follows:

(2.3) $P(\bar{E}_M, n, N(C))$

$$\leq \binom{\binom{n}{2}}{N(C)}^{-1} \sum_{M < s < n - 2n^{-1}N(C)} \binom{n}{s} \binom{\binom{n}{2} - s(n-s)}{N(C)}.$$

For sufficiently large n we have

(2.4) $\binom{\binom{n}{2}}{N(C)}^{-1} \binom{n}{s} \binom{\binom{n}{2} - s(n-s)}{N(C)} < \begin{cases} e^s/s! & \text{if } s < n/\log n, \\ \\ e^{-s} & \text{if } s \geq n/\log n. \end{cases}$

The estimates and inequality (2.3) give the relation

$$\lim_{n \to \infty} P(\bar{E}_{\log\log n}, n, N(C)) = 0.$$

We now claim that

(2.5) $P(\bar{\mathfrak{A}} \cap E_{\log\log n}, n, N(C))$

$$\leq \sum_{s=2}^{\log\log n} \binom{n}{s} \binom{\binom{n}{2}}{N(C)}^{-1} \sum_{r=1}^{\binom{s}{2}} \binom{\binom{s}{2}}{r} \binom{\binom{n-s}{2}}{N(C)-r}.$$

Indeed, if the largest component of a graph $\Gamma_{n,N(C)}$ has $n-s$ vertices and r is the number of edges in the graph with s vertices constituted by the remaining connected components of $\Gamma_{n,N(C)}$ then $r \geq 1$ and there exist $\binom{\binom{s}{2}}{r}$ ways to choose the edges, whereas the remaining $N(C) - r$ edges of the largest component can be chosen in $\binom{\binom{n-s}{2}}{N(C)-r}$ ways. It is not difficult to see that, for $s \geq 2$,

$$\sum_{r=1}^{\binom{s}{2}} \binom{\binom{s}{2}}{r} \binom{\binom{n-s}{2}}{N(C)-r} \leq N(C) \, 2^{\binom{s}{2}} \binom{n-s}{2}^{-1} \binom{\binom{n}{2} - s(n-s)}{N(C)}.$$

Combining (2.4) and (2.5) we obtain

$$P(\bar{\mathfrak{A}} \cap E_{\log\log n}, n, N(C)) \leq A \, n^{-1} \, (\log n)^{\log\log n + 1},$$

where A is an absolute constant. Hence it follows that

$$\lim_{n \to \infty} P(\bar{\mathfrak{A}} \cap E_{\log\log n}, n, N(C)) = 0.$$

Now the statement of the lemma is a corollary of the preceding estimates and the obvious equality

$$\mathbf{P}(\bar{\mathfrak{A}}) = \mathbf{P}(\bar{\mathfrak{A}} \cap E_M) + \mathbf{P}(\bar{E}_M) \, \mathbf{P}(\bar{\mathfrak{A}} \mid \bar{E}_M). \qquad \square$$

From the basic lemma one can deduce a number of theorems first proved by Érdös and Rényi.

Theorem 2.1 *Let $P_0(n, N(C))$ denote the probability of the connectedness of a random graph $\Gamma_{n,\,N(C)}$. Then*

$$(2.6) \qquad \lim_{n \to \infty} P_0(n, N(C)) = \exp\left\{ -e^{-2C} \right\}.$$

Proof Let $\mathfrak{B}$ be the event that a randomly chosen graph $\Gamma_{n,\,N(C)}$ has no isolated vertices. By applying the inclusion–exclusion principle one can deduce the explicit expression for the probability that $\mathfrak{B}$ occurs:

$$P(\mathfrak{B}, n, N(C)) = \binom{\binom{n}{2}}{N(C)}^{-1} \sum_{k=0}^{n} (-1)^k \binom{n}{k} \binom{\binom{n-k}{2}}{N(C)}.$$

Using Bonferroni's inequalities and taking into account the relation

$$(2.7) \qquad \lim_{n \to \infty} \binom{n}{k} \binom{\binom{n-k}{2}}{N(C)} \binom{\binom{n}{2}}{N(C)}^{-1} = \frac{1}{k!} \exp\left\{ -2\,k\,C \right\},$$

valid for any fixed k, we obtain

$$\lim_{n \to \infty} P(\mathfrak{B}, n, N(C)) = \exp\left\{ -e^{-2C} \right\}.$$

Now Theorem 2.1 follows from the obvious estimates

$$0 \le P(\mathfrak{B}, n, N(C)) - P_0(n, N(C)) \le P(\bar{\mathfrak{A}}, n, N(C)). \qquad \square$$

Theorem 2.2 *Let $P_k(n, N(C))$ be the probability that the largest component of a random graph $\Gamma_{n,\,N(C)}$ has $n - k$ vertices. Then*

$$(2.8) \qquad \lim_{n \to \infty} P_k(n, N(C)) = \frac{(e^{-2C})^k}{k!} \exp\left\{ -e^{-2C} \right\},$$

that is, the distribution of the number of vertices not belonging to the largest component of $\Gamma_{n,\,N(C)}$ is asymptotically Poisson with parameter $\lambda = e^{-2C}$.

Proof According to the basic lemma it is sufficient to consider only graphs of type $\mathfrak{A}$. Thus,

$$P_k(n, N(C)) = \binom{n}{k} \binom{\binom{n-k}{2}}{N(C)} \binom{\binom{n}{2}}{N(C)}^{-1} P_0(n-k, N(C)) + o(1), \quad n \to \infty.$$

Direct calculations show that, for any fixed k,

$$\lim_{n \to \infty} \frac{N(C) - \frac{1}{2}\,(n-k)\log(n-k)}{n-k} = C.$$

Putting this together with (2.6) and (2.7) we complete the proof of Theorem 2.2. $\qquad \square$

Theorem 2.3 *Let* $\xi(n, N(C))$ *be the number of connected components in a random graph* $\Gamma_{n,\,N(C)}$*. Then*

(2.9) $$\lim_{n \to \infty} \mathbf{P}\left\{\xi(n, N(C)) = k + 1\right\} = \frac{(e^{-2C})^k}{k!}\exp\left\{-e^{-2C}\right\}$$

for any $k = 0, 1, 2, \ldots$.

Proof As in the preceding theorem we may restrict ourselves to considering graphs of type $\mathfrak{A}$ and owing to this fact Theorem 2.3 is an obvious corollary to Theorem 2.2. □

Now our starting point is a graph of n isolated vertices and we consider the following random procedure for joining some of these vertices. At each step a randomly chosen pair of vertices is joined by an edge. Any edge (or, what amounts to the same, any pair of vertices) has the same probability of being chosen at a given step if it has not been chosen before. The procedure terminates at the moment when the graph obtained becomes connected. The number of steps in the process is a random variable denoted by ν_n.

Theorem 2.4 *If* $n \to \infty$ *and* $|l| = O(n)$ *then*

(2.10) $$\mathbf{P}\left\{\nu_n = \left[\tfrac{1}{2}n\log n\right] + l\right\} = \frac{2}{n}\exp\left\{-\frac{2l}{n} - e^{-2l/n}\right\}\left(1 + o(1)\right)$$

and, for any $x \in (-\infty, \infty)$,

(2.11) $$\lim_{n \to \infty} \mathbf{P}\left\{\frac{\nu_n - \tfrac{1}{2}n\log n}{n} < x\right\} = \exp\left\{-e^{-2x}\right\}.$$

Proof It is clear that if

$$\nu_n = \left[\tfrac{1}{2}n\log n\right] + l = N + 1$$

then at the preceding step we have constructed an unconnected graph $\Gamma_{n,\,N}$, which may be connected by adding an edge. According to the basic lemma we may assume that $\Gamma_{n,\,N}$ consists of one component with $n - 1$ vertices and an isolated vertex. Since the last edge may be chosen in $n - 1$ ways from the set of $\binom{n}{2} - N$ remaining edges, we have

$$\mathbf{P}\{\nu_n = N + 1\} = (n - 1)\left(\binom{n}{2} - N\right)^{-1}\left(P_1(n, N) + o(1)\right).$$

Hence, applying Theorem 2.2, we obtain (2.10), which gives

$$\mathbf{P}\left\{\frac{\nu_n - \tfrac{1}{2}n\log n}{n} < x\right\} = \sum_{l < nx}\frac{2}{n}\exp\left\{-\frac{2l}{n} - e^{-2l/n}\right\}\left(1 + o(1)\right).$$

The sum on the right-hand side of the equality may be viewed as a Riemann sum for the corresponding integral and the probability in question is asymptotically equal to

$$\int\limits_{-\infty}^{2x} \exp\left\{-u-e^{-u}\right\} du = \exp\left\{-e^{-2x}\right\}.$$

Thus, Theorem 2.4 is proved. □

Let $P_n^{(r)}$ be the probability that r transpositions, chosen at random and equiprobably from the set of all transpositions of n elements, generate $\mathbb{S}_n$, the symmetric group of order n. From Theorem 2.1 one can deduce the following useful fact proved by Dénes and Török [27].

Theorem 2.5 *If*

$$r = \left[\tfrac{1}{2}n\log n + Cn\right],$$

where C is a fixed constant, then

$$\lim_{n\to\infty} P_n^{(r)} = \exp\left\{-e^{-2C}\right\}.$$

Proof With each random tuple of transpositions $R_n^{(r)} = \{\tau_1, \tau_2, \ldots, \tau_r\}$ we associate the corresponding random Pólya graph $\Gamma(R_n^{(r)})$ in which an edge joins vertices i and j if and only if $\tau_s = (i,j) \in R_n^{(r)}$. It is known that $R_n^{(r)}$ generates $\mathbb{S}_n$ if and only if the graph $\Gamma(R_n^{(r)})$ is connected. This proves Theorem 2.5. □

An obvious corollary to Theorem 2.5 is the following relation:

$$\lim_{n\to\infty} P_0(n,N) = \begin{cases} 1 & \text{if } \dfrac{N - \tfrac{1}{2}n\log n}{n} \to +\infty, \\[2mm] 0 & \text{if } \dfrac{N - \tfrac{1}{2}n\log n}{n} \to -\infty. \end{cases}$$

We denote by $\nu(n,N)$ the valency of a fixed vertex of a random graph $\Gamma_{n,N}$, that is, the number of edges incident with the vertex. The following theorem is proved in [32].

Theorem 2.6 *If*

$$N = \tfrac{1}{2}n\log n + \tfrac{1}{2}rn\log\log n + \alpha n + o(n),$$

where α is a real number and r is a nonnegative integer, then, for any $k = 0, 1, 2, \ldots$,

$$\lim_{n\to\infty} \mathbf{P}\left\{\nu(n,N) = k\right\} = \frac{\lambda^k}{k!}\, e^{-\lambda},$$

where $\lambda = e^{-2\alpha}/r!$.

The paper cited also contains a result concerning the distribution of the minimal valency $\tilde{\nu}(n,N)$ of vertices of a random graph which, in a sense, characterizes its strength of connectedness.

Theorem 2.7 *If*

$$N = \tfrac{1}{2} n \log n + r n \log\log n + \alpha n + o(n), \qquad n \to \infty,$$

then

$$\lim_{n\to\infty} \mathbf{P}\{\nu(n, N) = r\} = 1 - \exp\{-e^{-2\alpha}/r!\}.$$

Theorems 2.6 and 2.7 can be proved by the method used for Theorems 2.1–2.5.

We give one more result from the paper [32]. A graph Γ is said to have a *factor of the first degree* if there exists a subset S of edges in Γ such that each vertex of Γ is incident with exactly one edge belonging to S. Clearly, for a factor of the first degree to exist in a graph Γ it is necessary for the number of vertices in Γ to be even.

Theorem 2.8 *Let $n = 2m$ and*

$$N = \tfrac{1}{2} n \log n + \omega(n)\, n, \qquad \lim_{n\to\infty} \omega(n) = \infty,$$

and let $\widetilde{P}_{n,N}$ denote the probability that a random graph $\Gamma_{n,N}$ has a factor of the first degree. Then

$$\lim_{n\to\infty} \widetilde{P}_{n,N} = 1.$$

To prove the theorem one can employ the ideas used in the proofs of the preceding theorems in combination with the following well-known result of Tutte [118].

Theorem 2.9 *A graph Γ has a factor of the first degree if and only if the deletion of any r vertices of Γ $(r = 0, 1, \dots)$ and the edges incident with them leads to a graph Γ^* in which the number of connected components having an even number of vertices is less than $r + 1$.*

6.2.2   Evolution of random graphs

The study of random graphs was initiated by the paper of Érdös and Rényi [31], in which the authors investigated the change in structure of a random graph $\Gamma_{n,N}$ when $n \to \infty$ and N grows with n in one or another way. This process of changing the structure of a graph is called its *evolution*. After the appearance of the pioneering work by Érdös and Rényi [31], the study of evolving graphs became popular among specialists in combinatorial probability. A detailed description of the corresponding results is given in the monograph by Bollobás [12], whose works, together with the results of other authors, enriched this field of combinatorial probability considerably [10, 11, 13, 44]. For this reason we discuss here only some basic notions and facts concerning the evolution of random graphs and in the main have used the paper [31]. To describe the evolution more rigorously it is convenient to use the notion of threshold function.

Let A be some property of random graphs and $P_{n,N}(A)$ be the probability that a graph $\Gamma_{n,N}$, chosen at random with probability $\left(\binom{n}{2}\atop N\right)^{-1}$, possesses this property. A function $A(n)$ is called a *threshold function* of property A if $A(n)$ tends monotonically to infinity as $n \to \infty$ and

$$\lim_{n\to\infty} P_{n,N}(A) = \begin{cases} 0, & N/A(n) \to 0, \\ 1, & N/A(n) \to \infty. \end{cases}$$

A function $A(n)$ is said to be a *regular threshold function* of property A if there exists a function $F(x)$, called the *threshold distribution function* of A, such that

$$\lim P_{n,N}(A) = F(x) \qquad \text{as} \qquad N/A(n) \to x \in (0,\infty),$$

for each continuity point of the function $F(x)$.

The ratio $2N/n$ is called the *global degree* of the graph $\Gamma_{n,N}$. A graph $\Gamma_{n,N}$ is said to be *balanced* if it has no subgraph whose global degree exceeds $2N/n$.

Denote by J the property that a random graph consists of isolated vertices and isolated edges. As there are exactly $\binom{n}{2N}(2N)!/(2^N N!)$ ways of constructing such graphs, the probability of $\Gamma_{n,N}$ having this property is

$$\mathbf{P}_{n,N}(J) = \frac{(n)_{2N}}{2^N\left[\binom{n}{2}\right]_N} = 1 + O\left(\frac{N^2}{n}\right).$$

Hence it follows that the graph $\Gamma_{n,N}$ consists of isolated vertices and isolated edges with probability close to 1, while $N = o(n^{1/2})$ and $n \to \infty$. Thus, the probability that trees or cycles with three or more vertices appear in $\Gamma_{n,N}$ is close to 0 for large n. The threshold functions for some properties of connected components of a random graph are described by the following theorem.

Theorem 2.10 *Let* $k \geq 2$, $k-1 \leq l \leq \binom{k}{2}$ *and let* $B_{k,l}$ *be a nonempty class of balanced graphs with* k *vertices and* l *edges. The function* $A(n) = n^{2-k/l}$ *is a threshold function of the property that a random graph* $\Gamma_{n,N}$ *has at least one subgraph isomorphic to an element of the class* $B_{k,l}$.

Corollary 1 *A threshold function of the property that a random graph* $\Gamma_{n,N}$ *contains a tree with* k *vertices as a subgraph is*

$$A(n) = n^{(k-2)/(k-1)}, \qquad k \geq 3.$$

Corollary 2 *A threshold function of the property that a random graph* $\Gamma_{n,N}$ *has a cycle with* k *vertices,* $k \geq 3$, *is* $A(n) = n$.

Corollary 3   *A threshold function of the property that a random graph* $\Gamma_{n,N}$ *contains a complete subgraph with* k *vertices is*

$$A(n) = n^{2(k-2)/(k-1)}, \qquad\qquad k \geq 3.$$

Here we give only a sketch of the proof of Theorem 2.10. Let $B_{kl}^{(n)}$ be the set of all subgraphs of a complete graph with n vertices that are isomorphic to some elements B_{kl} and let b_{kl} be the number of elements in B_{kl}. Denote by $\zeta_{n,N}$ the number of subgraphs $\gamma \in B_{kl}^{(n)}$ isomorphic to a subgraph of a randomly chosen graph $\Gamma_{n,N}$. It is clear that

$$\zeta_{n,N} = \sum_{\gamma \in B_{kl}^{(n)}} \delta(\gamma),$$

where $\delta(\gamma) = 1$ if γ is a subgraph of $\Gamma_{n,N}$ and $\delta(\gamma) = 0$ otherwise. The expectation of $\zeta_{n,N}$ is

$$\mathbf{E}\,\zeta_{n,N} = \binom{k}{l} b_{kl} \binom{\binom{n}{2} - l}{N - l} \binom{\binom{n}{2}}{N}^{-1}.$$

Hence it follows that for any k and l the asymptotic relation

$$\mathbf{E}\,\zeta_{n,N} \sim b_{kl}\,\frac{2\,N^l}{n^{2l-k}} \cdot \frac{1}{k!}$$

is valid as $n \to \infty$.

Thus, if $N = o(n^{2-k/l})$ as $n \to \infty$ then $\Gamma_{n,N}$ does not contain subgraphs from $B_{kl}^{(n)}$ with probability close to 1.

Assume now that $N = \omega(n)\,n^{2-k/l}$, where $\omega(n) \to \infty$ as $n \to \infty$. One can show that $\mathrm{Var}\,\zeta_{n,N}$ satisfies

$$\mathrm{Var}\,\zeta_{n,N} = O\left(\frac{(\mathbf{E}\,\zeta_{n,N})^2}{\omega(n)}\right).$$

Applying Chebyshev's inequality we find

$$\mathbf{P}\left\{\zeta_{n,N} \leq \tfrac{1}{2}\mathbf{E}\,\zeta_{n,N}\right\} = O\left(\frac{1}{\omega(n)}\right).$$

Clearly, $\mathbf{E}\,\zeta_{n,N} \to \infty$ as $\omega(n) \to \infty$ and, therefore, the probability of the event that $\Gamma_{n,N}$ includes at least one subgraph isomorphic to some element of B_{kl} tends to 1.

We now find the threshold distribution function for the trees with a given number of vertices. Denote by τ_k the number of isolated trees with k vertices in a random graph $\Gamma_{n,N}$.

Theorem 2.11 *If $n \to \infty$ and*

$$N \, n^{-(k-2)(k-1)} \longrightarrow \rho > 0$$

then the distribution of the random variable τ_k converges to a Poisson distribution with parameter $\lambda = (2\rho)^{k-1} k^{k-2}/k!$.

Observe that, under the conditions of the theorem, the probability of the appearance of nonisolated trees in $\Gamma_{n,N}$ tends to 1 as $n \to \infty$ (see Corollary 1 to Theorem 2.10).

The mean of τ_k can be calculated by the formula

$$\mathbf{E}\,\tau_k \;=\; E_k(n,N)\left(1 + O\!\left(\frac{N}{n^2}\right)\right),$$

where

$$E_k := E_k(n,N) = n\,\frac{k^{k-2}}{k!}\left(\frac{2N}{n}\right)^{k-1} e^{-2kN/n}.$$

We now consider the case where $\rho \to \infty$ as $n \to \infty$. Having stipulated this condition $\mathbf{E}\,\tau_k$ tends to infinity and the following theorem is valid.

Theorem 2.12 *If $n \to \infty$,*

$$N \, n^{-(k-2)(k-1)} \to \infty$$

and

$$\frac{N}{n} - \frac{1}{2k}\log n - \frac{k-1}{2k}\log\log n \longrightarrow -\infty,$$

then the distribution of the random variable $(\tau_k - E_k)\,E_k^{-1/2}$ converges to the standard normal distribution.

It follows from the asymptotic representation of $\mathbf{E}\,\tau_k$ that if

$$\frac{1}{n}\,N(n) - \frac{1}{2k}\log n - \frac{k-1}{2k}\log\log n \longrightarrow y \in (-\infty,\infty)$$

as $n \to \infty$ then $\mathbf{E}\,\tau_k \to \lambda$, where

$$\lambda = \frac{e^{-2ky}}{k \cdot k!}.$$

Under these conditions one can prove the following statement.

Theorem 2.13 *As $n \to \infty$, the distribution of τ_k converges to a Poisson distribution with parameter $\lambda = e^{-2ky}/(k \cdot k!)$.*

Further evolution of random graphs leads to the appearance of cycles whose probability distribution is described by the theorem below.

Theorem 2.14 *If $N \sim Cn$, $C > 0$, as $n \to \infty$ then the distribution of γ_k, the number of cycles of order k in $\Gamma_{n,N}$, converges to a Poisson distribution with parameter*

$$\lambda = \frac{(2C)^k}{2k} .$$

Observe that, under the same conditions, the distribution of the random variable γ_k^*, the number of isolated cycles of order k in $\Gamma_{n,N}$ converges, as $n \to \infty$, to a Poisson distribution with parameter

$$\mu = \frac{(2Ce^{-2C})^k}{2k} .$$

It should be noted that there exists no threshold function for isolated cycles since the function $1 - \exp\{-\mu\}$ attains its maximum $1 - \exp\{-e^{-k}/(2k)\}$ at $N(n) \sim Cn$, $C > 0$, and then begins to decrease.

One can also prove that under the conditions of Theorem 2.14 the random variable δ_k, the number of components in $\Gamma_{n,N}$ containing k vertices and k edges, has asymptotically a Poisson distribution with parameter

$$\nu = \frac{(2Ce^{-2C})^k}{2k} \sum_{j=0}^{k-3} \frac{k^j}{j!} .$$

Now we proceed to study the global properties of $\Gamma_{n,N}$ as $n \to \infty$. Denote by T the event that $\Gamma_{n,N}$ is a union of trees with no common vertices, and by $\bar{T}$ the complementary event, that is, the event that $\Gamma_{n,N}$ has at least one cycle. Let $P_{n,N}(T)$ and $P_{n,N}(\bar{T})$ be the probabilities of these events.

Theorem 2.15 *If $N = o(n)$ then $P_{n,N}(T)$ tends to 1 as $n \to \infty$.*

This theorem follows immediately from the obvious estimate

$$P_{n,N}(\bar{T}) \leq \sum_{k=3}^{n} \binom{n}{k} (k-1)! \binom{\binom{n}{2} - k}{N - k} \binom{\binom{n}{2}}{N}^{-1} = O\left(\frac{N}{n}\right).$$

Let $V_{n,N} = V(\Gamma_{n,N})$ be the number of vertices belonging to the isolated trees of a random graph $\Gamma_{n,N}$. The following theorem is valid.

Theorem 2.16 *If $Nn^{-1} \to C > 0$ as $n \to \infty$ then the mean number of vertices belonging to the isolated trees of a random graph $\Gamma_{n,N}$ satisfies the relation*

$$\lim_{n \to \infty} \mathbf{E}\,\frac{V_{n,N}}{n} = \begin{cases} 1 & \text{if } C \leq \tfrac{1}{2}, \\[2mm] x(C)/(2C) & \text{if } C > \tfrac{1}{2}, \end{cases}$$

where

$$x(C) := \sum_{k=1}^{\infty} \frac{k^{k-1}}{k!} \, (2 \, C \, e^{-2C})^k.$$

We see from the theorem that the structure of $\Gamma_{n,\,N}$ changes drastically when the ratio N/n passes through the point $C = \frac{1}{2}$. The next two theorems refine the character of the evolution.

Theorem 2.17 *Let* $n \to \infty$ *and* $N \, n^{-1} \to C$. *Then, for any sequence* ω_n, $n = 1, 2, \ldots$, *tending to infinity with* n *arbitrarily slowly, the following statements are valid:*

(a) *If* $C < \frac{1}{2}$ *then*

$$\lim_{n \to \infty} \mathbf{P} \, \{V_{n,\,N} \geq n - \omega_n\} = 1 \,.$$

(b) *If* $C > \frac{1}{2}$ *and* x *is the unique solution of the equation*

$$x \, e^{-x} = 2 \, N \, \exp \{ -2N \, n^{-1} \}$$

satisfying the condition $0 < x < 1$, *then*

$$\lim_{n \to \infty} \mathbf{P} \left\{ \left| V_{n,\,N} - \frac{n^2 \, x}{2 \, N} \right| > \omega_n \, \sqrt{n} \right\} = 0 \,.$$

Let $V_{n,\,N}(r)$ be the number of vertices in $\Gamma_{n,\,N}$ belonging to the trees of the graph, each of which has at least r vertices, and let $\tau_{n,N}(r)$ be the total number of such vertices in isolated trees.

Theorem 2.18 *If* $n \to \infty$ *and* $N \, n^{-1} \to \frac{1}{2}$ *then, for any* $\delta > 0$,

$$\mathbf{P} \left\{ \left| \frac{V_{n,\,N}(r)}{n} - \sum_{k=r}^{\infty} \frac{k^{k-1}}{k!} \, e^{-k} \right| < \delta \right\} \longrightarrow 1 \,,$$

$$\mathbf{P} \left\{ \left| \frac{\tau_{n,\,N}(r)}{n} - \sum_{k=r}^{\infty} \frac{k^{k-2}}{k!} \, e^{-k} \right| < \delta \right\} \longrightarrow 1 \,.$$

Denote by $H_{n,\,N}$ the number of cycles in $\Gamma_{n,\,N}$. One can show that if $N \sim C \, n$, $C \in [0, \frac{1}{2}]$, as $n \to \infty$ then

$$\lim_{n \to \infty} \mathbf{E} \, H_{n,\,N} = -\tfrac{1}{2} \log \, (1 - 2 \, C) - C - C^2, \qquad C < \tfrac{1}{2} \,,$$

$$\mathbf{E} \, H_{n,\,N} = \tfrac{1}{4} \log n \, (1 + o(1)) \,, \qquad\qquad\qquad C = \tfrac{1}{2} \,.$$

The second formula, an estimate for $\operatorname{Var} H_{n,\,N}$, and Chebyshev's inequality yield that, in the case $C = \frac{1}{2}$,

$$\lim_{n \to \infty} \mathbf{P}\left\{ \left| \frac{H_{n,\,N}}{\log n} - \frac{1}{4} \right| < \varepsilon \right\} = 1$$

for any $\varepsilon > 0$.

If K is the event that a random graph $\Gamma_{n,\,N}$ has at least one cycle and $N \sim C\,n$, $C \in [0, \frac{1}{2}]$, as $n \to \infty$, then

$$\lim_{n \to \infty} P_{n,\,N}(K) = 1 - \exp\{C + C^2\}\sqrt{1 - 2C};$$

in particular,

$$P_{n,\,N}(K) \to 1 \quad \text{for} \quad C = \tfrac{1}{2} \quad \text{and} \quad n \to \infty.$$

Denote by $H^*_{n,\,N}$ the number of vertices in a cycle of a random graph $\Gamma_{n,\,N}$. If $N n^{-1} \to C \in (0, \frac{1}{2})$ as $n \to \infty$ then the following relation is valid:

$$\lim_{n \to \infty} \mathbf{E}\,H^*_{n,\,N} = \frac{4\,C^3}{1 - 2C}.$$

Under the same conditions the mean value of the random variable $\nu_{n,\,N}$, the number of vertices in a component of $\Gamma_{n,\,N}$ having one cycle, is given as $n \to \infty$ by the formulae

$$\lim_{n \to \infty} \mathbf{E}\,\nu_{n,\,N} = \frac{1}{2} \sum_{k=3}^{\infty} (2\,C\,e^{-2C})^k \sum_{j=0}^{k-3} \frac{k^j}{j!}, \qquad C \neq \tfrac{1}{2},$$

$$\mathbf{E}\,\nu_{n,\,N} = \frac{1}{12} \Gamma\!\left(\frac{1}{3}\right) n^{2/3}\left(1 + o(1)\right), \qquad C = \tfrac{1}{2}.$$

Here Γ is the gamma-function as before.

The following theorem gives a rather visual picture of the structure of evolving random graphs.

Theorem 2.19 *If $N \sim C\,n$, $C \in (0, \frac{1}{2})$, as $n \to \infty$ then, with probability tending to 1, all the connected components of a random graph $\Gamma_{n,\,N}$ are either trees or contain precisely one cycle.*

Proof Denote by $\psi_{n,\,N}$ the number of vertices in $\Gamma_{n,\,N}$ that belong to the components, each of which has at most n vertices and for which the number of vertices is less than the number of edges. The mean value of $\psi_{n,\,N}$ can be estimated as follows:

$$\mathbf{E}\,\psi_{n,\,N} \leq \sum_{k=4}^{[\sqrt{\log n}]} k \binom{n}{k} 2^{\binom{k}{2}} \binom{\binom{n-k}{2}}{N-k-1} \binom{\binom{n}{2}}{N}^{-1} = O\left(n^{\frac{1}{2}\,\log 2 - 1}\right),$$

from which it follows that

$$\mathbf{P}\{\psi_{n,N} \geq 1\} = O\left(n^{\frac{1}{2}\log 2 - 1}\right).$$

This proves Theorem 2.19 since, according to Theorem 2.17, the probability of the existence of a component of $\Gamma_{n,N}$ not being a tree with more than $(\log n)^{1/2}$ vertices tends to 0.

Additional information concerning the structure of $\Gamma_{n,N}$ under the condition $N \sim Cn$ as $n \to \infty$ is given by the mean of the number $\xi_{n,N}$ of components in $\Gamma_{n,N}$:

$$\mathbf{E}\,\xi_{n,N} = \begin{cases} n - N + O(1) & \text{if} \quad 0 < C < \frac{1}{2}, \\[2mm] n - N + O(\log n) & \text{if} \quad C = \frac{1}{2}, \\[2mm] \dfrac{n}{4C}\left(2x(C) - x^2(C)\right) & \text{if} \quad C > \frac{1}{2}, \end{cases}$$

where

$$x(C) := \sum_{k=1}^{\infty} \frac{k^{k-1}}{k!} \, (2\,C\,e^{-2C})^k.$$

Hence, applying Chebyshev's inequality, we obtain

$$\mathbf{P}\{|\xi_{n,N} - n + N| < \omega_n\} \longrightarrow 1, \qquad\qquad 0 < C < \tfrac{1}{2},$$

$$\mathbf{P}\left\{\left|\frac{\xi_{n,N}}{n} - \frac{1}{4C}\left(2x(C) - x^2(C)\right)\right| < \varepsilon\right\} \longrightarrow 1, \qquad C \geq \tfrac{1}{2},$$

where $\varepsilon > 0$ and ω_n is a function tending to infinity arbitrarily slowly as $n \to \infty$.

Let $\Delta_{n,N}$ be the number of vertices in a tree of maximal size which is a component of $\Gamma_{n,N}$. We put

$$\alpha = 2C - 1 - \log 2C, \qquad \widetilde{\Delta}_{n,N} = \Delta_{n,N} - \alpha^{-1}\left(\log n - \tfrac{5}{2}\log\log n\right)$$

and assume that $N \sim Cn$, $C \neq \tfrac{1}{2}$, and $\omega_n \to \infty$ arbitrarily slowly as $n \to \infty$. The following relation is valid:

$$\lim_{n\to\infty} \mathbf{P}\{-\omega_n \leq \widetilde{\Delta}_{n,N} \leq \omega_n\} = 1.$$

Under the same conditions we have, for $C = \tfrac{1}{2}$,

$$\lim_{n\to\infty} \mathbf{P}\{\Delta_{n,N} \geq \omega_n\, n^{2/3}\} = 0, \qquad \lim_{n\to\infty} \mathbf{P}\{\Delta_{n,N} \geq \omega_n^{-1}\, n^{2/3}\} = 1.$$

One can find a more systematic description of the evolution of random graphs with detailed proofs in [10, 31].

6.2.3 Other models of random graphs

We refer to one more class of random graphs constructed as follows. Let a set of vertices, say $V_n = \{1, 2, \ldots, n\}$, be given. Consider a stochastic procedure according to which each of the $\binom{n}{2}$ possible edges occurs with the same probability p, $0 < p < 1$, independently of all other edges. As a result we obtain a random graph $\Gamma_n(p)$. Another equivalent way of constructing $\Gamma_n(p)$ is to remove, with probability $q = 1 - p$, each of the $\binom{n}{2}$ edges of a complete graph with the set of vertices V_n. The evolution of such random graphs as $n \to \infty$ is studied in considerable detail in Bollobás' monograph [10], where the author has established the connection of the results describing the structure of $\Gamma_n(p)$ under various restrictions relating n and p with the results of the preceding subsection. For this reason, here we concern ourselves with only two problems related to this subject.

Denote by P_n the probability of the connectedness of the graph $\Gamma_n(p)$. It can easily be seen that the following recurrence relation is valid:

$$P_n = 1 - \sum_{j=1}^{n-1} \binom{n-1}{j-1} P_j \, q^{j\,(n-j)}.$$

From this relation we deduce that, for any fixed $q \in (0, 1)$,

$$P_n = 1 - n\,q^{n-1} + O\big((n\,q^n)^2\big)$$

as $n \to \infty$, and thus

$$\lim_{n \to \infty} P_n = 1.$$

We now use another procedure for constructing a random graph $G_n(p)$ with n vertices. Initially, we have n isolated vertices and then, with probability p, each pair of vertices is joined by an edge independently of other pairs of vertices. We denote by $\bar{G}_n(p)$ the graph which is complementary to $G_n(p)$. A pair of vertices in $\bar{G}_n(p)$ is joined by an edge if and only if the pair has no common edge in $G_n(p)$.

Let $v_n(p, k)$ be the number of complete subgraphs with k vertices in $G_n(p)$. Direct calculations show that

$$\mathbf{E}\, v_n(p, k) = \binom{n}{k} \, p^{\binom{k}{2}}.$$

Substituting q for p here, we obtain the corresponding expression for the average number of complete subgraphs with k vertices in the random graph $\bar{G}_n(p)$:

$$\mathbf{E}\, v_n(q, k) = \binom{n}{k} \, q^{\binom{k}{2}}.$$

Choosing $k = k(n)$ to satisfy the inequality

$$k > \max\left\{\left[\frac{2\log n}{\log(1/p)}\right], \left[\frac{2\log n}{\log(1/q)}\right]\right\},$$

where $[z]$ is the integral part of z, we see that, in this case,

$$\mathbf{E}\, v_n(p,k) = o(1), \qquad \mathbf{E}\, v_n(q,k) = o(1), \qquad n \to \infty.$$

Using these estimates one can prove the following statement.

Theorem 2.20 *For each fixed $p \in (0,1)$ and for any $n > n_0(p)$ there exists a graph G_n with n vertices which contains no complete subgraph with more than $2\log n/\log(1/p)$ vertices and whose complementary graph $\bar{G}_n$ contains no complete subgraph with more than $2\log n/\log(1/q)$ vertices, $p + q = 1$.*

Putting $p = q = 1/2$ we obtain, as a corollary to the theorem, a result proved by Érdös and Turán [30].

Theorem 2.21 *For each sufficiently large n there exists a graph G_n with n vertices such that neither G_n nor $\bar{G}_n$ contains a complete subgraph with more than $2\log n/\log 2$ vertices.*

Theorems 2.20 and 2.21 are examples of such existence theorems in combinatorics in the proof of which one uses methods of probability theory rather than explicit constructions. One can find other examples showing how methods of probability theory can be used to prove combinatorial existence theorems in the excellent book [37].

6.3   Random mappings

6.3.1   Distribution of the number of cyclic vertices

We consider the uniform distribution on the set of all mappings of an n-element set $\mathbf{X}$ onto itself and study the random variable ζ_n, the number of cyclic vertices in the graph $\Gamma(\mathbf{X}, \sigma)$ of a random mapping $\sigma \in \mathfrak{G}_n$. From formula (0.13) it follows that

$$(3.1) \qquad \mathbf{P}\{\zeta_n = k\} = \frac{k\,(n)_k}{n^{k+1}}, \qquad k = 0, 1, \ldots, n.$$

The following theorem is due to Harris [54], where we use the notation $u(k) := k/\sqrt{n}$ for $k = 0, 1, 2, \ldots$.

Theorem 3.1 *If $n \to \infty$ and $u(k) = o(n^{1/6})$ then*

$$(3.2) \qquad \mathbf{P}\left\{\frac{\zeta_n}{\sqrt{n}} = \frac{k}{\sqrt{n}}\right\} = u(k)\,\exp\{-u^2(k)/2\}\,\frac{1}{\sqrt{n}}\left(1 + o(1)\right).$$

The proof of the theorem coincides with that of Theorem 1.4 from Section 1 of this chapter.

Of special interest is the generating function of ζ_n:

$$(3.3) \qquad f_n(x) = \sum_{k=1}^{n} \frac{k\,(n)_k}{n^{k+1}}\, x^k.$$

After simple transforms we have

$$(3.4) \qquad f_n(x) = 1 + (x-1)\,(n-1)!\sum_{k=0}^{n-1} \frac{1}{k!}\left(\frac{x}{n}\right)^{n-k-1}.$$

By finding the first two derivatives of the function at the point $x = 1$ one can obtain the following expressions for the mean and variance of ζ_n:

$$(3.5) \qquad \begin{cases} \mathbf{E}\,\zeta_n = \dfrac{n!}{n^n}\displaystyle\sum_{k=0}^{n-1}\dfrac{n^k}{k!}\,, \\[4mm] \operatorname{Var}\zeta_n = 2\,n - \dfrac{n!}{n^n}\displaystyle\sum_{k=0}^{n-1}\dfrac{n^k}{k!} - \left(\dfrac{n!}{n^n}\displaystyle\sum_{k=0}^{n-1}\dfrac{n^k}{k!}\right)^{2}. \end{cases}$$

Hence, using Stirling's formula and equality (1.20), we deduce

$$(3.6) \qquad \mathbf{E}\,\zeta_n = \sqrt{\frac{\pi\,n}{2}}\,\left(1 + o(1)\right)$$

and

$$(3.7) \qquad \operatorname{Var}\zeta_n = \left(2 - \frac{\pi}{2}\right) n \left(1 + o(1)\right).$$

6.3.2 Distribution of the number of components

We denote by $\varkappa_n$ the number of connected components of the graph $\Gamma(\mathbf{X}, \sigma)$ of a random mapping $\sigma \in \mathfrak{S}_n$. From (0.14) we know the distribution of $\varkappa_n$:

$$(3.8) \qquad \mathbf{P}\{\varkappa_n = j\} = \sum_{k=j}^{n}\binom{n-1}{k-1}\frac{|s(k,j)|}{n^k}\,, \qquad j = 1,\,2,\,\ldots,\,n.$$

Multiplying both sides of equality (3.8) by z^k and summing the relations obtained over k we find the generating function of $\varkappa_n$:

$$(3.9) \qquad g_n(z) = \sum_{k=1}^{n}\binom{z+k-1}{k}\frac{k\,(n)_k}{n^{k+1}}\,.$$

Lemma *The function $g_n(z)$ has the following asymptotic representation as $n \to \infty$:*

$$(3.10) \qquad g_n(z) = \frac{\Gamma(\frac{1}{2}(z+1))}{\Gamma(z)}\,(2\,n)^{(z-1)/2}\left(1 + o(1)\right),$$

where $o(1) \to 0$, as $n \to \infty$, uniformly for $z \in [1 - \delta,\, 1 + \delta]$, $0 < \delta < 1$.

Proof We represent $g_n(z)$ as follows:

$$(3.11) \qquad g_n(z) = H_1 + H_2 + H_3 = \sum_{k=1}^{n^{1/2-\epsilon}} + \sum_{k=n^{1/2-\epsilon}}^{n^{1/2+\epsilon}} + \sum_{k=n^{1/2+\epsilon}}^{n},$$

where $\delta\,(2\,(\delta+2))^{-1} < \epsilon < \frac{1}{6}$.

The first summand H_1 admits the estimate

$$(3.12) \qquad H_1 \le \frac{z\,e^{-z}}{n} \sum_{k=1}^{n^{1/2-\epsilon}} k^z = O\left(n^{\delta\,(1/2-\epsilon)-2\,\epsilon}\right),$$

which is uniform for $z \in [1 - \delta, 1 + \delta]$, $0 < \delta < 1$. In a similar way one can deduce the following estimate for H_3:

$$(3.13) \qquad H_3 \le \frac{1}{n\,\Gamma(z)} \sum_{k=n^{1/2+\epsilon}}^{n} k^z \exp\left\{-\frac{1}{n}\binom{k}{2}\right\} \left(1 + o(1)\right)$$

$$= O\left(n^{1+\delta} \exp\left\{-\tfrac{1}{2}\,n^{2\,\epsilon}\right\}\right),$$

which is uniform for z from the specified domain.

The sum H_2, which gives the main contribution to $g_n(z)$, can be represented in the form

$$H_2 = \frac{1}{n\,\Gamma(z)} \sum_{k=n^{1/2-\epsilon}}^{n^{1/2+\epsilon}} k^z\, e^{-k^2/(2n)} \left(1 + o(1)\right).$$

Hence we find

$$H_2 = \frac{n^{(z-1)/2}}{\Gamma(z)} \int_{n^{-\epsilon}}^{n^{\epsilon}} x^z\, e^{-x^2/2}\, dx \left(1 + o(1)\right).$$

Setting $y = x^2/2$ in the preceding integral and writing the gamma-function as

$$\Gamma(z) = \int_0^{\infty} y^{z-1}\, e^{-y}\, dy$$

we obtain

$$(3.14) \qquad H_2 = \frac{\Gamma(\frac{1}{2}\,(z+1))}{\Gamma(z)}\,(2\,n)^{(z-1)/2} \left(1 + o(1)\right),$$

where $o(1) \to 0$ as $n \to \infty$ uniformly for z from the domain indicated. The lemma now follows from representation (3.11) and estimates (3.12)–(3.14). $\qquad\qquad \square$

Let us study the asymptotic behavior of the random variable $\varkappa_n$ as $n \to \infty$. Put

$$\varkappa'_n := \left(\varkappa_n - \tfrac{1}{2} \log n\right) \Big/ \sqrt{\tfrac{1}{2} \log n}\,.$$

Theorem 3.2 ([113]) *If* $n \to \infty$ *then the distribution of the random variable* $\varkappa_n$ *converges to the standard normal distribution.*

Proof According to the lemma just proved, as $n \to \infty$, the moment generating function $G_n(t)$ of the random variable $\varkappa'_n$ has the asymptotic representation

$$G_n(t) = g_n(e^{t/\sigma}) \exp\left\{ -\frac{t}{2}\,\frac{\log n}{\sigma} \right\}$$

$$= \frac{\Gamma(\tfrac{1}{2}(e^{t/\sigma}+1))}{\Gamma(e^{t/\sigma})}\, e^{t^2/2}\left(1+o(1)\right),$$

where $\sigma^2 = \tfrac{1}{2}\log n$ and $o(1) \to 0$, as $n \to \infty$, uniformly for $t \in [-\delta', \delta']$, $\delta' > 0$. Hence we deduce

$$\lim_{n\to\infty} G_n(t) = e^{t^2/2}, \qquad t \in [-\delta', \delta']\,.$$

Theorem 3.2 now follows from Curtiss' theorem. □

By differentiating the generating function $g_n(z)$ at the point $z = 1$ one can find the mean and variance of $\varkappa_n$:

$$(3.15) \qquad \mathbf{E}\,\varkappa_n = \sum_{j=1}^{n} \frac{(n)_j}{j\, n^j}\,,$$

$$(3.16) \qquad \mathrm{Var}\,\varkappa_n = 2 \sum_{1 \le i < j \le n} \frac{(n)_j}{i\, j\, n^j} + \sum_{j=1}^{n} \frac{(n)_j}{j\, n^j} - \left(\sum_{j=1}^{n} \frac{(n)_j}{j\, n^j} \right)^2.$$

Using as $n \to \infty$ the estimates

$$\sum_{j=1}^{\sqrt{n}} \frac{(n)_j}{j\, n^j} = \int_{1/\sqrt{n}}^{1} \frac{1}{x}\, e^{-x^2/2}\, dx \left(1+o(1)\right) = \frac{1}{2} \log n \left(1+o(1)\right),$$

$$\sum_{j=\sqrt{n}+1}^{n} \frac{(n)_j}{j\, n^j} = O(1),$$

we conclude that

$$(3.17) \qquad \mathbf{E}\,\varkappa_n = \tfrac{1}{2} \log n \left(1+o(1)\right).$$

Similar arguments show that

$$(3.18) \qquad \mathrm{Var}\,\varkappa_n = \tfrac{1}{2} \log n \left(1+o(1)\right).$$

6.3.3 Distribution of the number of images
and inverse images of an element

Let σ be a mapping chosen at random and equiprobably from the set of mappings of an n-element set $\mathbf{X}$ onto itself. For a fixed element $x \in \mathbf{X}$, the set

$$S_\sigma(x) := \left\{ x, \sigma(x), \sigma^2(x), \ldots, \sigma^{n-1}(x) \right\}$$

is called the *set of images* of the element x with respect to σ.

We denote by $s_\sigma(x)$ the number of distinct elements in $S_\sigma(x)$. An element $y \in \mathbf{X}$ is called an *inverse image* of x with respect to a mapping σ if there exists a nonnegative integer k such that $\sigma^k(y) = x$. The number of such elements $y \in \mathbf{X}$ is denoted by $p_\sigma(x)$. Let $l_\sigma(x)$ be the length of the cycle in the component of the graph $\Gamma(\mathbf{X}, \sigma)$ containing x. Setting $s_n := s_\sigma(x)$ and $l_n := l_\sigma(x)$ we conclude that if $1 \leq j \leq k \leq n$ then

$$(3.19) \qquad \mathbf{P}\left\{ s_n = k, l_n = j \right\} = \binom{n-1}{k-1} (k-1)! \, \frac{n^{n-k}}{n^n} = \frac{(n)_k}{n^{k+1}}.$$

Hence by summing we find

$$(3.20) \qquad \mathbf{P}\left\{ s_n = k \right\} = \frac{k\,(n)_k}{n^{k+1}}, \qquad\qquad k = 0, 1, \ldots, n,$$

$$(3.21) \qquad \mathbf{P}\left\{ l_n = j \right\} = \sum_{k=j}^{n} \frac{(n)_k}{n^{k+1}}, \qquad\qquad j = 1, 2, \ldots, n.$$

Comparing formulae (3.1) and (3.20) and taking Theorem 3.1 into account we see that the following theorem is valid.

Theorem 3.3 *The number of images of an element with respect to a random mapping $\sigma \in \mathfrak{S}_n$ has, as $n \to \infty$, the limiting distribution coinciding with the limiting distribution of the number of cyclic elements: if $u(k) = k/\sqrt{n} = o(n^{1/6})$, $k = 1, 2, \ldots$, then*

$$\mathbf{P}\left\{ \frac{s_n}{\sqrt{n}} = \frac{k}{\sqrt{n}} \right\} = u(k) \, \exp\left\{ -u^2(k)/2 \right\} \frac{1}{\sqrt{n}} \left(1 + o(1) \right).$$

This theorem may also be proved by another method. Indeed, applying Stirling's formula to the right-hand side of (3.19) one can establish the relation

$$\mathbf{P}\left\{ \frac{s_n}{\sqrt{n}} = \frac{k}{\sqrt{n}}, \ \frac{l_n}{\sqrt{n}} = \frac{j}{\sqrt{n}} \right\} = \exp\left\{ -u^2(k)/2 \right\} \frac{1}{n} \left(1 + o(1) \right),$$

where $j \leq k$ and $u(k) = o(n^{1/6})$. It follows from this relation that the asymptotic density of the random vector $(s_n/\sqrt{n}, l_n/\sqrt{n})$ is

$$f(x,y) = e^{-x^2/2}, \qquad 0 < y \leq x < \infty.$$

By integrating the density we find the limiting densities of the random variables $s_n/\sqrt{n}$ and $l_n/\sqrt{n}$ respectively (see [53]):

$$f_1(u) = u\, e^{-u^2/2}, \qquad u > 0,$$

$$f_2(y) = \sqrt{2\pi}\,(1 - \Phi(y)), \qquad y > 0,$$

where

$$\Phi(y) := \frac{1}{\sqrt{2\pi}} \int_{-\infty}^{y} e^{-u^2/2}\, du.$$

We denote by τ_n the number of inverse images of a fixed element $x \in \mathbf{X}$ with respect to a randomly chosen mapping $\sigma \in \mathfrak{S}_n$. We claim that

$$(3.22) \quad \mathbf{P}\{\tau_n = k\} = \binom{n}{k} \frac{k^{k-1}}{n^k} \left(1 - \frac{k}{n}\right)^{n-k}, \qquad k = 1,2,\ldots,n-1.$$

Indeed, one may choose $k - 1$ inverse images of x other than x in $\binom{n-1}{k-1}$ ways. Each group of chosen elements generates k^{k-2} various trees with root x and with edges directed to the root. The number of mappings of the set of the remaining $n - k$ elements into itself is $(n - k)^{n-k}$. The element x may have any of n elements as its image. Therefore,

$$(3.23) \quad \mathbf{P}\{\tau_n = k\} = \binom{n-1}{k-1} \frac{k^{k-2}(n-k)^{n-k}\, n}{n^n}.$$

Obvious transforms lead to (3.22), the exact distribution of the random variable τ_n (see [90]).

Theorem 3.4 *The number of inverse images of a fixed element with respect to a random equiprobable mapping $\sigma \in \mathfrak{S}_n$ has, as $n \to \infty$, the limiting distribution*

$$\lim_{n\to\infty} \mathbf{P}\{\tau_n = k\} = \frac{k^{k-1}}{e^k\, k!}, \qquad k = 1,2,\ldots.$$

This theorem follows from (3.23) by direct application of Stirling's formula.

6.3.4 Distribution of the height and order of elements

We consider the symmetric semigroup $\mathfrak{G}_n$ of all mappings of an n-element set $\mathbf{X}$ onto itself. Let $x \in \mathbf{X}$ be a vertex in the graph $\Gamma(\mathbf{X}, \sigma)$ of a mapping $\sigma \in \mathfrak{G}_n$. The length of the path leading from x to the nearest cyclic point of $\Gamma(\mathbf{X}, \sigma)$ is called the *height* of x with respect to σ and will be denoted by $h(x, \sigma)$. By definition, the height of a cyclic element is equal to zero. The quantity $h(\sigma) := \max_{x \in \mathbf{X}} h(\sigma, x)$ is called the *height of the mapping* $\sigma \in \mathfrak{G}_n$. The following theorem is valid [88].

Theorem 3.5 *Let $h(\sigma)$ be the height of a mapping σ chosen at random and equiprobably from among the elements of the symmetric semigroup $\mathfrak{G}_n$. Then*

$$\lim_{n \to \infty} \mathbf{P}\left\{ h(\sigma)\,(2\,n)^{-1/2} < x \right\} = \sum_{k=-\infty}^{\infty} (-1)^k\, e^{-k^2 x^2}.$$

The number of distinct elements of the sequence $(\sigma, \sigma^2, \dots)$, $\sigma \in \mathfrak{G}_n$, is called the *order* of σ in the semigroup $\mathfrak{G}_n$ and is denoted by $Q(\sigma)$. One can define the order of σ as the minimal positive integer $Q(\sigma)$ such that $\sigma^q = \sigma^{Q(\sigma)+1}$ for some q, $0 < q \le Q(\sigma)$. If $C_\sigma \subseteq \mathbf{X}$ is the set of cyclic elements of σ then the restriction of the mapping σ to C_σ is called the *skeleton* of σ and is denoted by σ^*. It is clear that

$$Q(\sigma) = Q(\sigma^*) + \max\left\{ 0,\, h(\sigma) - 1 \right\}.$$

The following theorem is due to Harris [54].

Theorem 3.6 *If $\sigma \in \mathfrak{G}_n$ is a random equiprobable mapping then the distribution of the random variable*

$$\frac{\log Q(\sigma) - 8^{-1} \log^2 n}{24^{-1/2}\, \log^{3/2} n}$$

converges, as $n \to \infty$, to the standard normal distribution.

The proof of the theorem is based on the following important result, which is due to Érdös and Turán [38].

Theorem 3.7 *Let $Q(s)$ be the order of the element s chosen at random and equiprobably from the set $\mathbb{S}_n$, the symmetric group of permutations of order n. Then the distribution of the random variable*

$$\frac{\log Q(s) - 2^{-1} \log^2 n}{3^{-1/2}\, \log^{3/2} n}$$

is, as $n \to \infty$, asymptotically normal with parameters $(0, 1)$.

One can also prove that if $G(n) = \{\max Q(s),\ s \in \mathbb{S}_n\}$ then

$$\lim_{n \to \infty} \frac{\log G(n)}{\sqrt{n \log n}} = 1.$$

Moreover, if $W(n)$ is the number of distinct values taken by $Q(s)$, $s \in \mathbb{S}_n$, then

$$W(n) = \exp\left\{ \sqrt{\frac{4\pi^2 n}{6 \log n}} \left(1 + o(1)\right) \right\}, \qquad n \to \infty,$$

and among the values taken by $Q(s)$ all but the number of values of order $o(W(n))$ have the form

$$\exp\left\{ \frac{\sqrt{6} \log 2}{\pi} \sqrt{n \log n} \left(1 + o(1)\right) \right\}.$$

6.4   Random mappings with bounded height

6.4.1   The number of mappings with bounded height

Let $\mathfrak{G}_n^h$ denote the family of all mappings σ of an n-element set $\mathbf{X}$ onto itself, with the underlying condition that the mapping height does not exceed h, that is, the heights of trees of the graphs $\Gamma(\mathbf{X}, \sigma)$, $\sigma \in \mathfrak{G}_n^h$, do not exceed h. An element $x \in \mathbf{X}$ is said to belong to the ith *layer* of a mapping σ if i is the least number such that an integer $p > 0$ exists for which $\sigma^i(x) = \sigma^{i+p}(x)$. Clearly, the ith layer consists of elements of height i of the mapping σ and, in particular, the zeroth layer consists of the cyclic elements of σ. With every mapping $\sigma \in \mathfrak{G}_n^h$ we associate an $(h+1)$-dimensional vector $\mathbf{K} = (k, k_1, \ldots, k_h)$, $k + k_1 + \cdots + k_h = n$, where k_i is the number of elements $x \in \mathbf{X}$ in the ith layer of the mapping σ and $k = k_0$. The vector $\mathbf{K}$ is called the *layer diagram* of the mapping $\sigma \in \mathfrak{G}_n^h$.

Lemma 1 *The number* $T(k, k_1, \ldots, k_h)$ *of mappings* $\sigma \in \mathfrak{G}_n^h$ *with layer diagram* $\mathbf{K} = (k, k_1, \ldots, k_h)$, $k + k_1 + \cdots + k_h = n$, *is equal to*

$$(4.1) \qquad T(k, k_1, \ldots, k_h) = \frac{n!}{k_1! \cdots k_h!}\, k^{k_1} k_1^{k_2} \cdots k_{h-1}^{k_h},$$

where $0^0 = 1$ *and* $k_{i+1} = k_{i+2} = \cdots = k_h = 0$ *if* $k_i = 0$.

Proof Direct calculations show that the number of ways of distributing the elements among the layers is equal to $n!\,(k!\,k_1! \cdots k_h!)^{-1}$, while the number of mappings of the k_i elements of the ith layer onto the k_{i-1} elements of the $(i-1)$th layer is $k_{i-1}^{k_i}$, and the number of mappings of the cyclic elements onto themselves is $k!$.   □

Let $T_n^{(h)}(k)$ and $T_{nj}^{(h)}$ denote the numbers of mappings $g \in \mathfrak{G}_n^h$ with k cyclic elements and j components, respectively, and let $T_n^{(h)}$ be the total number of mappings in $\mathfrak{G}_n^h$. Then

$$(4.2) \qquad T_n^{(h)}(k) = \sum_{k_1 + \cdots + k_h = n-k} T(k, k_1, \ldots, k_h),$$

$$(4.3) \qquad T_{nj}^{(h)} = \sum_{k=j}^{n} \frac{1}{k!} C(k, j) \, T_n^{(h)}(k),$$

$$(4.4) \qquad T_n^{(h)} = \sum_{k=0}^{n} T_n^{(h)}(k),$$

where $C(k, j)$ is the number of permutations of order k with j cycles and where for convenience we set

$$T_0^{(h)}(0) = T_{00}^{(h)} = T_0^{(h)} = 1.$$

To investigate the asymptotic behavior of $T(k, k_1, \ldots, k_h)$, $k + k_1 + \cdots + k_h = n$, as $n \to \infty$, we first consider the following function of h independent variables $x, x_1, \ldots, x_{h-1}$ with $x + x_1 + \cdots + x_{h-1} + x_h = n$ and h parameters $l_1, l_2, \ldots, l_h$:

$$(4.5) \quad f_n(x, x_1, \ldots, x_{h-1}; \mathbf{L})$$

$$:= \frac{x^{x_1 + l_1} \, x_1^{x_2 + l_2} \cdots x_{h-2}^{x_{h-1} + l_{h-1}} \, x_{h-1}^{x_h + l_h}}{\Gamma(x_1 + 1) \, \Gamma(x_2 + 1) \cdots \Gamma(x_{h-1} + 1) \, \Gamma(x_h + 1)},$$

where $\mathbf{L} = (l_1, l_2, \ldots, l_h)$, $\Gamma(x)$ is the gamma-function and all the variables under consideration are nonnegative.

Lemma 2 *The function* $f_n(x, x_1, \ldots, x_{h-1}; \mathbf{L})$ *has, as* $n \to \infty$, *a unique maximum at the point* $(x^0, x_1^0, \ldots, x_h^0)$, *which can be specified by the conditions:*

$$(4.6) \quad \begin{cases} x^0 = r_0 \, n, \quad x_1^0 = r_1 \, n, \quad \ldots, \quad x_h^0 = r_h \, n, \\[4pt] r_0 + r_1 + \cdots + r_h = 1, \\[4pt] r_j = \dfrac{\rho_1 \rho_2 \cdots \rho_j}{1 + \rho_1 + \rho_1 \rho_2 + \cdots + \rho_1 \rho_2 \cdots \rho_h}, \quad j = 1, 2, \ldots, h, \end{cases}$$

where

$$(4.7) \quad \begin{cases} \rho_j = L_{h-j}(\rho), \quad\quad j = 1, 2, \ldots, h-1, \\[4pt] L_0(\rho) = \rho, \quad L_k(\rho) = \rho \exp\{L_{k-1}(\rho)\}, \quad k = 1, 2, \ldots, h, \end{cases}$$

and $\rho = \rho_h$ *is the unique real solution of the equation*

$$(4.8) \qquad L_h(\rho) = 1.$$

Proof Let an extremum of the function $f_n(x, x_1, \ldots, x_{h-1}; \mathbf{L})$ exist at some point $(x^0, x_1^0, \ldots, x_{h-1}^0)$. Then we have the system of equations:

$$(x_1^0 + l_1) / x^0 - \log x_{h-1}^0 + \gamma(x_h^0 + 1) = 0,$$

$$(x_2^0 + l_2) / x_1^0 + \log (x^0 / x_{h-1}^0) - \gamma(x_1^0 + 1) + \gamma(x_h^0 + 1) = 0,$$

$$\cdots\cdots\cdots\cdots\cdots\cdots\cdots\cdots\cdots\cdots\cdots\cdots$$

$$(x_{h-1}^0 + l_{h-1}) / x_{h-2}^0 + \log (x_{h-3}^0 / x_{h-1}^0) - \gamma(x_{h-2}^0 + 1) + \gamma(x_h^0 + 1) = 0,$$

$$(x_h^0 + l_h) / x_{h-1}^0 + \log (x_{h-2}^0 / x_{h-1}^0) - \gamma(x_{h-1}^0 + 1) + \gamma(x_h^0 + 1) = 0,$$

where $\gamma(x) = \Gamma'(x) / \Gamma(x)$ is the logarithmic derivative of the gamma-function. Being concerned with only the nonnegative solutions of this system as $n \to \infty$ and introducing the notation

$$\rho_1 = x_1^0 / x^0, \qquad \rho_2 = x_2^0 / x_1^0, \qquad \ldots,$$

$$\rho_{h-1} = x_{h-1}^0 / x_{h-2}^0, \qquad \rho_h = x_h^0 / x_{h-1}^0,$$

we have

(4.9) $$\begin{cases} \rho_h \, e^{\rho_1} = 1, & \rho_h \, e^{\rho_2} = \rho_1, \qquad \ldots, \\ \rho_h \, e^{\rho_{h-1}} = \rho_{h-2}, & \rho_h \, e^{\rho_h} = \rho_{h-1}. \end{cases}$$

Setting $\rho = \rho_h$ and solving system (4.9), we obtain the desired result. □

It is easy to see that the quantity ρ_h decreases monotonically in h and remains bounded from below. Consequently, the limit $\lim_{h \to \infty} \rho_h$ exists and from equation (4.8) it follows that

$$\lim_{h \to \infty} \rho_h = e^{-1}.$$

Computations show that for $h = 500$ the values of ρ_h and e^{-1} do not differ until the sixth decimal place.

Lemma 3 Let $\boldsymbol{\alpha} = (\alpha_0, \alpha_1, \ldots, \alpha_{h-1})$, $\alpha_0 + \alpha_1 + \cdots + \alpha_h = 0$, $\alpha_j = o(n^{1/6})$, $j = 0, 1, \ldots, h$, and put

$$\varphi_n(\boldsymbol{\alpha}, \mathbf{L}) := f_n(x^0 + \alpha_0\sqrt{n}, \, x_1^0 + \alpha_1\sqrt{n}, \ldots, x_{h-1}^0 + \alpha_{h-1}\sqrt{n}; \mathbf{L}).$$

Then, for any fixed $l_1, l_2, \ldots, l_h$, the following asymptotic representation is valid as $n \to \infty$:

(4.10) $$\varphi_n(\boldsymbol{\alpha}, \mathbf{L}) = n^{-h/2} \, \varphi_n(\mathbf{0}, \mathbf{L}) \exp\left\{ -\tfrac{1}{2} \, Q(\alpha_0, \alpha_1, \ldots, \alpha_h) \right\} (1 + o(1)),$$

where

$$\varphi_n(0,\mathbf{L}) = \frac{n^l}{(2\pi)^{h/2}\,\rho^n} \cdot \frac{r_0^{l_1}\,r_1^{l_2}\,\cdots\,r_{h-1}^{l_h}}{\sqrt{r_1\,r_2\,\cdots\,r_h}}, \qquad l = l_1 + l_2 + \cdots + l_h,$$

and

$$(4.11) \qquad Q(\alpha_0,\alpha_1,\ldots,\alpha_h) = \frac{r_1}{r_0^2}\,\alpha_0^2 + \sum_{j=1}^{h-1} \frac{r_{j+1}+r_j}{r_j^2}\,\alpha_j^2$$

$$+ \frac{\alpha_h^2}{r_h} - 2\sum_{j=0}^{h-1}\frac{1}{r_j}\,\alpha_j\alpha_{j+1}.$$

Proof We apply Stirling's formula for the gamma-function to estimate $\varphi_n(\boldsymbol{\alpha},\mathbf{L})$ as $n \to \infty$:

$$\varphi_n(\boldsymbol{\alpha},\mathbf{L}) = \frac{1}{(2\pi)^{h/2}} \prod_{j=1}^{h} \frac{\left(r_{j-1}\,n + \alpha_{j-1}\sqrt{n}\right)^{r_j n + \alpha_j \sqrt{n} + l_j}}{\left(r_j\,n + \alpha_j\sqrt{n}\right)^{r_j n + \alpha_j \sqrt{n} + 1/2}}$$

$$\times \exp\left\{ \sum_{j=1}^{h}\left(r_j\,n + \alpha_j\sqrt{n}\right) + O\!\left(\frac{1}{n}\right)\right\}.$$

Hence we deduce

$$\varphi_n(\boldsymbol{\alpha},\mathbf{L}) = \frac{n^l}{(2\pi n)^{h/2}} \cdot \frac{r_0^{l_1}\,r_1^{l_2}\,\cdots\,r_{h-1}^{l_h}}{\sqrt{r_1\,r_2\,\cdots\,r_h}}$$

$$\times \exp\left\{(1-r_0)\,n - \alpha_0\sqrt{n} + O\!\left(\frac{1}{n}\right)\right\} \prod_{j=1}^{h}\left(\frac{r_{j-1}}{r_j}\right)^{r_j n + \alpha_j \sqrt{n}}$$

$$\times \prod_{j=1}^{h} \frac{\left(1+\alpha_{j-1}/(\sqrt{n}\,r_{j-1})\right)^{r_j n + \alpha_j \sqrt{n} + l_j}}{\left(1+\alpha_j/(\sqrt{n}\,r_j)\right)^{r_j n + \alpha_j \sqrt{n} + 1/2}}.$$

Raising the ith equation in system (4.9) to the power r_{i-1} and multiplying the resulting equations, we obtain

$$\rho_1^{r_1}\,\rho_2^{r_2}\,\cdots\,\rho_h^{r_h} = \rho_h \exp\left\{r_1 + r_2 + \cdots + r_h\right\}.$$

Now, using the preceding relation and the equalities $r_{j-1}/r_j = 1/\rho_j$, $j = 1,2,\ldots,h$, we find that

$$\exp\left\{(1-r_0)\,n\right\} \prod_{j=1}^{h}\left(\frac{r_{j-1}}{r_j}\right)^{r_j n} = \frac{1}{\rho^n}.$$

By expanding the logarithm $\log\left(1+\alpha_{j-1}/(r_{j-1}\sqrt{n})\right)$ in a Maclaurin series in the domain $\alpha_j = o(n^{1/6})$, $j = 0, 1, \ldots, h$, we see that

$$\prod_{j=1}^{h}\left(1+\frac{\alpha_{j-1}}{r_{j-1}\sqrt{n}}\right)^{r_j n + \alpha_j \sqrt{n} + l_j}$$

$$= \exp\left\{\sqrt{n}\sum_{i=1}^{h}\frac{r_i}{r_{i-1}}\alpha_{i-1} + \sum_{i=1}^{h}\frac{\alpha_{i-1}\alpha_i}{r_{i-1}} - \frac{1}{2}\sum_{i=1}^{h}\frac{r_i}{r_{i-1}^2}\alpha_{i-1}^2 + O\left(\frac{\bar\alpha^3}{\sqrt{n}}\right)\right\},$$

where $\bar\alpha = \max_{1\le j\le h}\alpha_j$. Similarly,

$$\prod_{j=1}^{h}\left(1+\frac{\alpha_j}{r_j\sqrt{n}}\right)^{r_j n + \alpha_j\sqrt{n}+1/2} = \exp\left\{-\alpha_0\sqrt{n}+\frac{1}{2}\sum_{i=1}^{h}\frac{\alpha_i^2}{r_i}+O\left(\frac{\bar\alpha^3}{\sqrt{n}}\right)\right\}.$$

Further, we note that

$$\exp\left\{\sum_{i=1}^{h}\rho_i\alpha_{i-1}\right\}\prod_{i=1}^{h}\left(\frac{1}{\rho_i}\right)^{\alpha_i} = (\rho_h e^{\rho_1})^{\alpha_0}\prod_{i=1}^{h-1}\left(\frac{\rho_h}{\rho_i}e^{\rho_{i+1}}\right)^{\alpha_i} = 1.$$

Collecting the above estimates we obtain the following asymptotic representation for $\varphi_n(\boldsymbol{\alpha}, \mathbf{L})$:

$$\varphi_n(\boldsymbol{\alpha}, \mathbf{L}) = \frac{n^l}{(2\pi n)^{h/2}\rho^n}\cdot\frac{r_0^{l_1}r_1^{l_2}\cdots r_{h-1}^{l_h}}{\sqrt{r_1 r_2\cdots r_h}}$$

$$\times\exp\left\{-\frac{1}{2}\sum_{i=1}^{h}\frac{\alpha_i^2}{r_i}-\frac{1}{2}\sum_{i=1}^{h}\frac{r_i}{r_{i-1}^2}\alpha_{i-1}^2+\sum_{i=1}^{h}\frac{\alpha_{i-1}\alpha_i}{r_{i-1}}+O\left(\frac{\bar\alpha^3}{\sqrt{n}}\right)\right\}.$$

Hence Lemma 3 follows.   □

Lemma 4   *The quadratic form* $Q(\alpha_0, \alpha_1, \ldots, \alpha_h)$, $\alpha_0 + \alpha_1 + \cdots + \alpha_h = 0$, *is positive semidefinite and has rank* h. *By the linear transform*

(4.12) $$\beta_k = \frac{r_{k+1}}{r_k^2}\alpha_k - \frac{1}{r_k}\alpha_{k+1}, \qquad k = 0, 1, \ldots, h-1,$$

the form $Q(\alpha_0, \alpha_1, \ldots, \alpha_h)$ *is converted to a sum of squares:*

(4.13) $$Q(\alpha_0, \alpha_1, \ldots, \alpha_h) = \sum_{k=0}^{h-1}\frac{r_k^2}{r_{k+1}}\beta_k^2.$$

If we regard $Q(\alpha_0, \alpha_1, \ldots, \alpha_h)$ *as the quadratic form* $\widehat{Q}(\alpha_0, \alpha_1, \ldots, \alpha_{h-1})$ *of the variables* $\alpha_0, \alpha_1, \ldots, \alpha_{h-1}$ *then this form is positive definite.*

Proof Let D_k be the kth principal minor of the matrix $Q(\alpha_0, \alpha_1, \dots, \alpha_h)$. Then

$$D_0 = 1, \qquad\qquad D_1 = \frac{r_1}{r_0^2},$$

$$D_k = \frac{r_{k-1} + r_k}{r_{k-1}^2} D_{k-1} - \frac{1}{r_{k-2}^2} D_{k-2}, \qquad k = 2, 3, \dots, h,$$

and, consequently,

(4.14) $$D_k = \frac{r_1 r_2 \cdots r_k}{(r_0 r_1 \cdots r_{k-1})^2}, \qquad k = 1, 2, \dots, h.$$

For the determinant D_{h+1} of the matrix under consideration we have

$$D_{h+1} = \frac{1}{r_h} D_h - \frac{1}{r_{h-1}^2} D_{h-1} = 0.$$

Applying Jacobi's method for converting a quadratic form to a sum of squares we get

$$Q(\alpha_0, \alpha_1, \dots, \alpha_h) = \sum_{k=0}^{h-1} \frac{D_k}{D_{k+1}} \beta_k^2, \qquad D_0 = 1,$$

and, therefore, relation (4.13) is proved. This implies that the quadratic form $\widehat{Q}(\alpha_0, \alpha_1, \dots, \alpha_{h-1})$ is nonnegative and nonsingular. Consequently, it is positive definite. □

Lemma 5 *The Jacobian J_h for the transform from the coordinates α_0, α_1, $\dots$, α_{h-1} to the coordinates β_0, β_1, $\dots$, β_{h-1} can be computed by the formula*

(4.15) $$J_h = \left| \left(\frac{\partial \alpha_i}{\partial \beta_j} \right)_0^{h-1} \right| = r_0^2\, r_1\, r_2 \cdots r_{h-1}, \qquad h = 2, 3, \dots,$$

with $J_1 = r_0^2$.

Proof Indeed, if $\boldsymbol{\alpha} = (\alpha_0, \alpha_1, \dots, \alpha_{h-1})$, $\boldsymbol{\beta} = (\beta_0, \beta_1, \dots, \beta_{h-1})$ and R is the matrix of a linear transform such that $\boldsymbol{\alpha} = R^{-1} \boldsymbol{\beta}$, then $J_h = \det R^{-1}$. From system (4.12) we see that the matrix R has the form

(4.16) $$R = \begin{pmatrix} \dfrac{r_1}{r_0^2} & -\dfrac{1}{r_0} & 0 & \cdots & 0 & 0 \\[2ex] 0 & \dfrac{r_2}{r_1^2} & -\dfrac{1}{r_1} & \cdots & 0 & 0 \\[2ex] \vdots & \vdots & \vdots & \ddots & \vdots & \vdots \\[2ex] 0 & 0 & 0 & \cdots & \dfrac{r_{h-1}}{r_{h-2}^2} & -\dfrac{1}{r_{h-2}} \\[2ex] \dfrac{1}{r_{h-1}} & \dfrac{1}{r_{h-1}} & \dfrac{1}{r_{h-1}} & \cdots & \dfrac{1}{r_{h-1}} & \dfrac{r_h}{r_{h-1}^2} + \dfrac{1}{r_{h-1}} \end{pmatrix}.$$

By decomposing $\det \mathrm{R}$ with respect to the last row, or by induction, one can show that

(4.17)
$$\det \mathrm{R} = \frac{1}{r_0^2 \, r_1 \cdots r_{h-1}}, \qquad h > 1,$$

and $\det \mathrm{R} = r_0^{-2}$ for $h = 1$. $\qquad\qquad\qquad\qquad\qquad\qquad\qquad$ □

Lemma 6 *Let a tuple* $\mathbf{L} = (l_1, l_2, \ldots, l_h)$ *be fixed and let*

(4.18) $\quad T_n^{(h)}(\mathbf{L}) := \displaystyle\sum_{k+k_1+\cdots+k_h=n} \frac{n!}{k_1! \, k_2! \cdots k_h!}$

$$\times \; k^{k_1+l_1} \, k_1^{k_2+l_2} \cdots k_{h-1}^{k_h+l_h}.$$

Then

(4.19) $\quad T_n^{(h)}(\mathbf{L}) = n! \, n^l \, \rho^{-n} \, r_0^{l_1+1} \, r_1^{l_2} \cdots r_{h-1}^{l_h} \, (1 + o(1)), \qquad n \to \infty.$

Proof We choose an arbitrary $\varepsilon \in (0, \tfrac{1}{6})$ and present $T_n^{(h)}(\mathbf{L})$ in the form

$$T_n^{(h)}(\mathbf{L}) = n! \, (S_1 + S_2),$$

where

$$S_1 := \sum_{|k_j - x_j^0| \le n^{2/3-\varepsilon}} f_n(k, k_1, \ldots, k_{h-1}; \mathbf{L}),$$

$$S_2 := \sum_{|k_j - x_j^0| > n^{2/3-\varepsilon}} f_n(k, k_1, \ldots, k_{h-1}; \mathbf{L}).$$

In the sum of S_1 it is understood that $|k_j - x_j^0| \le n^{2/3-\varepsilon}$ for all $j = 0, 1, \ldots, h-1$, where we agree to write $k_0 = k$ and $x_0^0 = x^0$, while in S_2 summation is taken over those $k, k_1, \ldots, k_h$ for which the inequality $|k_j - x_j^0| > n^{2/3-\varepsilon}$ holds for at least one j, $j = 0, 1, \ldots, h-1$.

From Lemma 3 it follows that

$$S_1 = f_n(x^0, x_1^0, \ldots, x_{h-1}^0; \mathbf{L}) \, I_h \, (1 + o(1)),$$

where

$$I_h := n^{-h/2} \sum_{\substack{|(k_j - x_j^0)\, n^{-1/2}| \le q(n,\varepsilon) \\ j=0,1,\ldots,h-1}}$$

$$\times \exp\left\{ -\tfrac{1}{2} Q\left((k_0 - x^0) n^{-1/2}, \ldots, (k_h - x_h^0) n^{-1/2} \right) \right\}$$

and $q(n, \varepsilon) := n^{1/6-\varepsilon}$.

We note that I_h is, in fact, a Riemann sum so that, as $n \to \infty$,

$$I_h = \int_{-q(n,\varepsilon)}^{q(n,\varepsilon)} \cdots \int_{-q(n,\varepsilon)}^{q(n,\varepsilon)} \exp\left\{ -\tfrac{1}{2} Q(\alpha_0, \alpha_1, \dots, \alpha_h) \right\} d\alpha_0 \, d\alpha_1 \, \dots \, d\alpha_{h-1}$$

$$\times \left(1 + o(1) \right).$$

Performing in the h-fold integral a change of variables by the linear transform of Lemma 4, we obtain

$$I_h = \int_{-C_0 \, q(n,\varepsilon)}^{C_0 \, q(n,\varepsilon)} \cdots \int_{-C_{h-1} \, q(n,\varepsilon)}^{C_{h-1} \, q(n,\varepsilon)} \exp\left\{ -\frac{1}{2} \sum_{k=0}^{h-1} \frac{r_k^2}{r_{k+1}} \beta_k^2 \right\}$$

$$\times J_h \, d\beta_0 \, d\beta_1 \, \dots \, d\beta_{h-1} \left(1 + o(1) \right),$$

where $C_0, C_1, \dots, C_{h-1}$ are some positive constants.

Recall now that, as $n \to \infty$,

$$\int_{-C_i \, q(n,\varepsilon)}^{C_i \, q(n,\varepsilon)} \exp\left\{ -\frac{1}{2} \frac{r_i^2}{r_{i+1}} \beta_i^2 \right\} d\beta_i = \frac{\sqrt{2\pi r_{i+1}}}{r_i} \left(1 + o(1) \right),$$

$$i = 0, 1, \dots, h - 1,$$

where the error term decreases exponentially. Applying Lemma 5 yields

$$I_h = (2\pi)^{h/2} \, r_0 \, \sqrt{r_1 r_2 \cdots r_h} \, \left(1 + o(1) \right).$$

Note that

$$\varphi_n(\mathbf{0}, \mathbf{L}) = f_n(x^0, x_1^0, \dots, x_{h-1}^0; \mathbf{L})$$

and, according to Lemma 3,

$$S_1 = n^l \, \rho^{-n} \, r_0^{l_1+1} \, r_1^{l_2} \, \cdots \, r_{h-1}^{l_h} \left(1 + o(1) \right).$$

For sufficiently large n and $\boldsymbol{\alpha} = (\alpha_0, \dots, \alpha_k, \dots, \alpha_{h-1})$ varying with n in such a way that $\alpha_j = o(n^{1/6-\varepsilon})$, $j \neq k$, and $\alpha_k \geq n^{1/6-\varepsilon}$, the following estimates are valid:

$$\varphi_n(\boldsymbol{\alpha}, \mathbf{L}) \leq n^{-h/2} \, \varphi_n(\mathbf{0}, \mathbf{L})$$

$$\times \exp\left\{ -\tfrac{1}{2} \widehat{Q}(\alpha_0, \dots, n^{1/6-\varepsilon}, \dots, \alpha_{h-1}) \right\} \left(1 + o(1) \right),$$

$$S_2 \leq 2^{h+1} \, \varphi_n(\mathbf{0}, \mathbf{L}) \, \gamma_1 \, n^h \, \exp\left\{ -\gamma_2 \, n^{1/3-2\varepsilon} \right\} J_h$$

$$\times \int_{-\infty}^{\infty} \cdots \int_{-\infty}^{\infty} \exp\left\{ -\frac{1}{2} \sum_{j=0}^{h-1} \frac{r_j^2}{r_{j+1}} \beta_j^2 \right\} d\beta_0 \, d\beta_1 \, \dots \, d\beta_{h-1},$$

where γ_1, γ_2 are some positive constants and $\varepsilon \in (0, \frac{1}{6})$ is arbitrary. Therefore, there exist positive constants C_1 and C_2 such that

$$S_2 \le C_1 \, n^{l+h} \, \rho^{-n} \, \exp\left\{ - C_2 \, n^{1/3 - 2\varepsilon} \right\}$$

for large n. This completes the proof of Lemma 6. □

Corollary 1 *If h is fixed, then, for the number of mappings of height not exceeding h, the asymptotic representation*

$$(4.20) \quad T_n^{(h)} = n! \, \rho^{-n} \, (1 + \rho_1 + \rho_1 \rho_2 + \cdots + \rho_1 \rho_2 \cdots \rho_h)^{-1} \, (1 + o(1))$$

is valid, as $n \to \infty$, where $\rho_1, \rho_2, \ldots, \rho_h$ are defined in Lemma 2 by equalities (4.7) and expressed in terms of the solution ρ of equation (4.8).

Proof For $\mathbf{L} = \mathbf{L}_0 = (0, 0, \ldots, 0)$ it follows from Lemma 6 that

$$(4.21) \qquad\qquad T_n^{(h)} = n! \, \rho^{-n} \, r_0 \, (1 + o(1)),$$

where, by Lemma 2,

$$(4.22) \qquad r_0 = (1 + \rho_1 + \rho_1 \rho_2 + \cdots + \rho_1 \rho_2 \cdots \rho_h)^{-1}. \qquad\qquad □$$

Corollary 2 *The number of mappings of height not exceeding 1 is equal to*

$$(4.23) \qquad\qquad T_n^{(1)} = \sum_{k=1}^{n} (n)_k \, k^{n-k}.$$

As $n \to \infty$, the asymptotic formula

$$(4.24) \qquad\qquad T_n^{(1)} = \frac{n!}{\tilde{\rho}^n \, (1 + \tilde{\rho})} \left(1 + o(1) \right)$$

is valid, where $\tilde{\rho} = 0.567 \ldots$ is the unique real solution of the equation

$$(4.25) \qquad\qquad \tilde{\rho} \, e^{\tilde{\rho}} = 1.$$

The corollary can easily be proved by applying the preceding results to $h = 1$.

6.4.2 The limiting distribution of the layer diagram of a random mapping

We consider the uniform distribution on $\mathfrak{G}_n^h$ and study the h-dimensional random variable $\boldsymbol{\xi} = (\xi_0, \xi_1, \ldots, \xi_{h-1})$, where ξ_i is the number of elements in the ith layer of a random mapping $\sigma \in \mathfrak{G}_n^h$. The distribution of the random vector $\boldsymbol{\xi}$, being the layer diagram of a random mapping from $\mathfrak{S}_n^h$, is as follows:

$$\mathbf{P}\{\xi_0 = k, \xi_1 = k_1, \ldots, \xi_{h-1} = k_{h-1}\}$$

$$= \frac{T(k, k_1, \ldots, k_h)}{T_n^{(h)}}, \qquad k + k_1 + \cdots + k_h = n.$$

The moments of $\boldsymbol{\xi}$ have the form

$$\mathbf{E}\,\xi_0^{l_1}\,\xi_1^{l_2}\cdots\xi_{h-1}^{l_h} = T_n^{(h)}(\mathbf{L})\,/\,T_n^{(h)}, \qquad \mathbf{L} = (l_1, l_2, \ldots, l_h)$$

and, by Lemma 6, the asymptotic representation

$$(4.26)\quad \mathbf{E}\,\xi_0^{l_1}\,\xi_1^{l_2}\cdots\xi_{h-1}^{l_h}$$

$$= r_0^{l_1}\,r_1^{l_2}\cdots r_{h-1}^{l_h}\,n^l\,(1+o(1)), \quad l = l_1 + l_2 + \cdots + l_h,$$

is valid as $n \to \infty$. In particular, the average number of points in the ith layer is

$$(4.27)\qquad \mathbf{E}\,\xi_i = r_i\,n\,(1+o(1)), \qquad i = 0, 1, \ldots, h,$$

where the coefficients $r_0, r_1, \ldots, r_h$ depend solely on h.

If $K_j := \mathbf{E}\xi_j/\mathbf{E}\xi_{j-1}$, $j = 1, 2, \ldots, h$, is the 'coefficient of relative dilution' of the jth layer with respect to the $(j-1)$th layer then, by the equality $r_j = \rho_j r_{j-1}$, we have

$$K_j = \rho_j, \qquad j = 1, 2, \ldots, h.$$

As $1 > K_1 > \cdots > K_h > 0$, we see that the relative impoverishment of neighboring layers increases with the increase in their indices. The following theorem gives more detailed information about the distribution of the number of points in the layers of a random mapping $\sigma \in \mathfrak{S}_n^h$.

Theorem 4.1 *If $(\xi_0, \xi_1, \ldots, \xi_{h-1})$ is the layer diagram of a random mapping $\sigma \in \mathfrak{S}_n^h$ then the distribution of the random vector*

$$\left(\,(\xi_0 - r_0\,n)/\sqrt{n},\,(\xi_1 - r_1 n)/\sqrt{n},\,\ldots,\,(\xi_{h-1} - r_{h-1})/\sqrt{n}\,\right)$$

converges, as $n \to \infty$, to a proper normal distribution with the covariance matrix $\Lambda = \widehat{Q}^{-1}$, where $\widehat{Q}$ is a matrix of the form $\widehat{Q}(\alpha_0, \alpha_1, \ldots, \alpha_{h-1})$, defined in Lemma 4, whose coefficients can be presented as functions of the root of equation (4.8).

Proof Lemmas 3 and 6 imply

$$\mathbf{P}\left\{(\xi_0 - x^0)/\sqrt{n} = \alpha_0,\ (\xi_1 - x_1^0)/\sqrt{n} = \alpha_1,\right.$$

$$\left.\dots,\ (\xi_{h-1} - x_{h-1}^0)/\sqrt{n} = \alpha_{h-1}\right\}$$

$$= \frac{n!}{T_n^{(h)}}\, f_n\left(x^0 + \alpha_0\sqrt{n},\ x_1^0 + \alpha_1\sqrt{n},\ \dots,\ x_{h-1}^0 + \alpha_{h-1}\sqrt{n};\ \mathbf{L}_0\right)$$

$$= \frac{\varphi_n(\mathbf{0}, \mathbf{L}_0)\, n!}{n^{h/2}\, T_n^{(h)}}\, \exp\left\{-\tfrac{1}{2}\widehat{Q}(\alpha_0, \alpha_1, \dots, \alpha_{h-1})\right\}\left(1 + o(1)\right).$$

Now $\widehat{Q}(\alpha_0, \alpha_1, \dots, \alpha_{h-1})$ being nonsingular guarantees the existence of the inverse quadratic form $\widehat{Q}^{-1}(\alpha_0, \alpha_1, \dots, \alpha_{h-1})$, whose matrix we denote by $\Lambda = \|\lambda_{ij}\|$, $i, j = 0, 1, \dots, h-1$. Since

$$\widehat{Q}(\alpha_0, \alpha_1, \dots, \alpha_{h-1}) = (\det \Lambda)^{-1} \sum_{i,\,k=0}^{h-1} \Lambda_{ik}\, \alpha_i\, \alpha_k,$$

where $\det \Lambda$ is the determinant of Λ and Λ_{ik} is the cofactor of the component λ_{ik}, we finally obtain

$$(4.28)\quad \mathbf{P}\left\{(\xi_0 - x^0)/\sqrt{n} = \alpha_0,\ (\xi_1 - x_1^0)/\sqrt{n} = \alpha_1,\right.$$

$$\left.\dots,\ (\xi_{h-1} - x_{h-1}^0)/\sqrt{n} = \alpha_{h-1}\right\}$$

$$= \frac{1}{(2\pi n)^{h/2}\,\sqrt{r_0^2\, r_1 \cdots r_h}}$$

$$\times \exp\left\{-\tfrac{1}{2}(\det \Lambda)^{-1} \sum_{i,\,k=0}^{h-1} \Lambda_{ik}\,\alpha_i\,\alpha_k\right\}\left(1 + o(1)\right).$$

If B is the diagonal matrix of the quadratic form of the variables $\beta_0, \beta_1, \dots,$ β_{h-1}, obtained from transforming $\widehat{Q}(\alpha_0, \alpha_1, \dots, \alpha_{h-1})$ to a sum of squares by the method specified in Lemma 4, then $\Lambda^{-1} = \mathrm{R}'\,\mathrm{B}\,\mathrm{R}$, where R is the same as in (4.16) and the prime signifies transposition. Now formula (4.17) implies that

$$\det \Lambda = r_0^2\, r_1 \cdots r_h.$$

It follows from equality (4.28) that the scaled random vector considered in the theorem has in the limit the probability density

$$(4.29)\quad \psi(\alpha_0, \alpha_1, \dots, \alpha_{h-1}) = \frac{1}{\sqrt{(2\pi)^h\, \det \Lambda}}$$

$$\times \exp\left\{-\tfrac{1}{2}(\det \Lambda)^{-1} \sum_{i,\,k=0}^{h-1} \Lambda_{ik}\,\alpha_i\,\alpha_k\right\}$$

and the characteristic function

$$(4.30) \qquad \Phi(t_0, t_1, \ldots, t_{h-1}) = \exp\left\{ -\frac{1}{2} \sum_{i,\,k=0}^{h-1} \lambda_{ik}\, t_i\, t_k \right\}. \qquad \square$$

6.4.3 The distribution of the number of components

We denote by η_{nh} the number of components in a random mapping $\sigma \in \mathfrak{G}_n^h$ and consider the generating function

$$a(z) := \sum_{j=1}^{n} \mathbf{P}\{\eta_{nh} = j\}\, z^j.$$

Lemma 7 *As $n \to \infty$, the generating function $a(z)$ has the following asymptotic representation:*

$$a(z) = \frac{(r_0\, n)^{z-1}}{\Gamma(z)}\left(1 + o(1)\right),$$

where $o(1) \to 0$ uniformly for z from some neighborhood $W = [1 - \delta, 1 + \delta]$, $0 < \delta < 1$, of the point $z = 1$.

Proof From Lemma 1 and formula (4.3) it follows that the generating function

$$A(z) := \sum_{j=1}^{n} T_{nj}^{(h)}\, z^j$$

is of the form

$$A(z) = n! \sum_{k=1}^{n} z\,(z+1)\,\cdots\,(z+k-1) \sum_{k_1 + \cdots + k_h = n-k} \frac{k^{k_1}\, k_1^{k_2} \cdots k_{h-1}^{k_h}}{k!\, k_1! \,\cdots\, k_h!}$$

or

$$A(z) = n! \sum_{k + k_1 + \cdots + k_h = n} F_n(k, k_1, \ldots, k_{h-1}; z),$$

where

$$F_n(x, x_1, \ldots, x_{h-1}; z) := \frac{x^{x_1}\, x_1^{x_2} \cdots x_{h-1}^{x_h}\, \Gamma(x + z)}{\Gamma(x+1)\,\Gamma(x_1 + 1) \,\cdots\, \Gamma(x_h + 1)\,\Gamma(z)}$$

and $x + x_1 + \cdots + x_h = n$.

It is easy to show that, for arbitrary z, $0 < z < \infty$, the function $F_n(x, x_1, \ldots, x_{h-1}; z)$ attains its maximum, as $n \to \infty$, at the same point $(x^0, x_1^0, \ldots, x_{h-1}^0; z)$ as the function $f_n(x, x_1, \ldots, x_{h-1}; z)$ involved in Lemma 2. Furthermore, as $n \to \infty$ and for $\alpha = (\alpha_0, \alpha_1, \ldots, \alpha_{h-1})$, $\alpha_0 + \alpha_1 + \cdots + \alpha_n = 0$, $\alpha_j = o(n^{1/6})$, $j = 0, 1, \ldots, h$, and $\mathbf{L}_0 = (0, 0, \ldots, 0)$, we have

$$F_n(x^0 + \alpha_0\sqrt{n}, \ldots, x^0_{h-1} + \alpha_{h-1}\sqrt{n}; z)$$

$$= \varphi_n(\boldsymbol{\alpha}, \mathbf{L}_0) \frac{\Gamma(x^0 + \alpha_0\sqrt{n} + z)}{\Gamma(x^0 + \alpha_0\sqrt{n} + 1)\,\Gamma(z)},$$

where $\varphi_n(\boldsymbol{\alpha}, \mathbf{L})$ is defined in Lemma 3. To estimate the gamma-function we apply Stirling's formula and obtain, as $n \to \infty$,

$$F_n(x^0 + \alpha_0\sqrt{n}, \ldots, x^0_{h-1} + \alpha_{h-1}\sqrt{n}; z)$$

$$= \frac{(r_0\,n)^{z-1}}{\Gamma(z)}\, f_n(x^0 + \alpha_0\sqrt{n}, \ldots, x^0_{h-1} + \alpha_{h-1}\sqrt{n}; \mathbf{L}_0)\left(1 + o(1)\right),$$

where $o(1) \to 0$ uniformly in a neighborhood W of the point $z = 1$. This equality implies that

$$A(z) = \frac{(r_0\,n)^{z-1}}{\Gamma(z)}\, n! \sum_{\boldsymbol{\alpha}} f_n(x^0 + \alpha_0\sqrt{n}, \ldots, x^0_{h-1} + \alpha_{h-1}\sqrt{n}; \mathbf{L}_0)\left(1 + o(1)\right),$$

where the sum represents an asymptotic expression for $T_n^{(h)}/n!$ as $n \to \infty$. Hence,

$$a(z) = \frac{A(z)}{A(1)} = \frac{(r_0\,n)^{z-1}}{\Gamma(z)}\left(1 + o(1)\right)$$

with $o(1)$ tending to zero uniformly for $z \in W$. $\qquad\square$

Theorem 4.2   *As $n \to \infty$, the distribution of the random variable*

$$\bar{\eta}_{nh} = \frac{\eta_{nh} - \log(r_0\,n)}{\sqrt{\log(r_0\,n)}}$$

is asymptotically normal with parameters $(0, 1)$.

Proof   From Lemma 7 we have, as $n \to \infty$, the following asymptotic representation for the moment generating function of η_{nh}:

$$b(z) = \frac{(r_0\,n)^{e^z - 1}}{\Gamma(e^z)}\left(1 + o(1)\right),$$

where $o(1)$ tends to zero uniformly in a neighborhood W' of the point $z = 0$, $W' = [-\delta', \delta']$, $\delta' > 0$. By Curtiss' theorem, to prove our theorem it suffices to show that for any fixed z and $\theta = \sqrt{\log(r_0\,n)}$

$$\lim_{n \to \infty} e^{-\theta z}\, \frac{(r_0\,n)^{u-1}}{\Gamma(u)} = e^{z^2/2}, \qquad u = e^{z/\theta}.$$

It is easy to justify this last equation. $\qquad\square$

We now consider the random variable $\varkappa_{nh}(r)$, the number of cycles of length r (or, in short, r-cycles) in the graph $\Gamma(\mathbf{X}, \sigma)$ of a random equiprobable mapping $\sigma \in \mathfrak{G}_n^h$. It is easy to write down the exact distribution of $\varkappa_{nh}(r)$ by the total probability formula:

$$(4.31) \qquad \mathbf{P}\{\varkappa_{nh}(r) = j\} = \frac{1}{T_n^{(h)}} \sum_{k=jr}^{n} \frac{1}{k!} C(k, j; r) T_n^{(h)}(k)$$

for $j = 1, 2, \ldots, [n/r]$, where $C(k, r; j)$ is the number of permutations of order k with j cycles of length r.

From the exact distribution of $\varkappa_{nh}(r)$ we deduce the following representation for the generating function of its factorial moments:

$$g_n(z) = \sum_{j=0}^{[n/r]} [M]_j \frac{z^j}{j!} = \sum_{k=0}^{n} \frac{T_n^{(h)}(k)}{T_n^{(h)}} \tilde{g}_k(z),$$

where

$$\tilde{g}_k(z) := \sum_{j=0}^{[k/r]} \frac{1}{r^j} \cdot \frac{z^j}{j!}$$

is the generating function of factorial moments of r-cycles of a random permutation of order k. Thus,

$$g_n(z) = \sum_{j=0}^{[n/r]} \sum_{k=rj}^{n} \frac{z^j}{r^j j!} \cdot \frac{T_n^{(h)}(k)}{T_n^{(h)}}.$$

We fix an arbitrary $\varepsilon > 0$ and let

$$\Delta = \left[(r_0 n - n^{2/3-\varepsilon})/r \right].$$

For $|z| < r$ we write $g_n(z) = S_1 + S_2$ by breaking up the outer summation into summation from 0 to Δ and from $\Delta + 1$ to $[n/r]$.

For the sum S_1 we have $rj \leq r_0 n - n^{2/3-\varepsilon}$; it follows therefore from the proof of Lemma 6 that, as $n \to \infty$,

$$\sum_{k=rj}^{n} \frac{T_n^{(h)}(k)}{T_n^{(h)}} = 1 + o(1).$$

Moreover,

$$S_2 = O\big(n/(\Delta+1)!\big) = o(1).$$

Hence, for $|z| < r$,

$$\lim_{n \to \infty} g_n(z) = \sum_{j=0}^{\infty} \frac{1}{r^j} \cdot \frac{z^j}{j!},$$

that is, the generating function of the factorial moments of $\varkappa_{nh}(r)$ converges, as $n \to \infty$, to the generating function of the factorial moments of a Poisson law. $\qquad \square$

Thus, we have just proved the following theorem.

Theorem 4.3 *As* $n \to \infty$, *the random variable* $\varkappa_{nh}(r)$ *is asymptotically distributed according to a Poisson law with parameter* $\lambda = r^{-1}$.

6.4.4 The distribution of the height and the number of images of a fixed element

The collection of various elements in $\{x, \sigma(x), \sigma^2(x), \dots, \sigma^{n-1}(x)\}$ represents the collection of images of an element $x \in \mathbf{X}$ under the mapping $\sigma \in \mathfrak{S}_n^h$. Let θ be the number of images of an element x under the mapping $\sigma \in \mathfrak{S}_n^h$, let v be the length of the cycle of $\Gamma(\mathbf{X}, \sigma)$ corresponding to the collection of images of x, and let $w = \theta - v$ be the height of x with respect to the mapping σ.

Lemma 8 *The joint distribution of the random variables v and w is given by the formula*

$$(4.32) \quad \mathbf{P}\{v = l, \, w = j\}$$

$$= \frac{1}{T_n^{(h)}} \sum_{k=l}^{n} \sum_{k_1 + \cdots + k_h = n-k} \frac{k_j}{k\,n} \cdot \frac{n!}{k_1! \cdots k_h!} \, k_1^{k_1} k_1^{k_2} \cdots k_{h-1}^{k_h}.$$

The exact distributions of v and w have the form

$$(4.33) \qquad \mathbf{P}\{v = l\} = \sum_{k=l}^{n} \frac{1}{k} \frac{T_n^{(h)}(k)}{T_n^{(h)}}, \qquad l = 1, 2, \dots, n,$$

$$(4.34) \qquad \mathbf{P}\{w = j\} = \frac{1}{n} \frac{T_n^{(h)}(\mathbf{L}_{j+1})}{T_n^{(h)}}, \qquad j = 0, 1, \dots, h-1,$$

where $T_n^{(h)}(\mathbf{L}_0) = T_n^{(h)}$, $\mathbf{L}_j = (0, \dots, 0, 1, 0, \dots, 0)$ (1 *appears in the jth position) and $T_n^{(h)}(\mathbf{L})$ can be calculated by* (4.18).

Proof The number of ways of choosing and ordering $l + j$ images of an element x is equal to $(n-1)_{l+j-1}$, while the number of mappings $\sigma \in \mathfrak{S}_n^h$ with specified images of x and a layer diagram $(k, k_1, \dots, k_h)$ is equal to

$$\frac{(n-l-j)!}{(k_1 - 1)! \cdots (k_j - 1)! \, k_{j+1}! \cdots k_h!}$$

$$\times \, k_1^{k_1 - 1} k_2^{k_2 - 1} \cdots k_j^{k_j - 1} k_j^{k_{j+1}} \cdots k_{h-1}^{k_h}.$$

Multiplying these values, summing the results and dividing the sum by $T_n^{(h)}$, we find the joint distribution of v and w. Hence, after simplification, we obtain (4.32). Marginal distributions of v and w are found by summing the joint distribution with respect to the corresponding parameters. $\qquad \square$

Theorem 4.4 *As* $n \to \infty$, *the height* w *of a fixed element* x *with respect to a random equiprobable mapping* $\sigma \in \mathfrak{S}_n^h$ *has the limiting distribution*

$$(4.35) \qquad \lim_{n \to \infty} \mathbf{P}\{w = j\} = r_j, \qquad\qquad j = 0, 1, \ldots, h,$$

where the probabilities $r_0, r_1, \ldots, r_h$ *are defined by formulae* (4.6). *The distribution of the length* v *of the cycle generated in the graph* $\Gamma(X, \sigma)$ *by the collection of images* x *is asymptotically concentrated on the interval* $[1, r_0 n]$ *and is uniform on that interval. More exactly,*

$$(4.36) \quad \mathbf{P}\{v = l\} \begin{cases} = \dfrac{1}{r_0 n}\left(1 + o(1)\right), & 1 \le l, \\[2mm] & r_0 n - n^{2/3-\varepsilon} \ge l, \\[2mm] \le \dfrac{1}{r_0 n}\left(1 + o(1)\right), & r_0 n - n^{2/3-\varepsilon} + 1 \le l, \\[2mm] & r_0 n + n^{2/3-\varepsilon} \ge l, \\[2mm] = o\left(\dfrac{1}{n}\right), & r_0 n + n^{2/3-\varepsilon} + 1 \le l, \\[2mm] & n \ge l. \end{cases}$$

Proof Distribution (4.35) can be calculated directly by the asymptotic formulae (4.19) and (4.21). As for the remaining statements of the theorem, we observe that

$$T_n^{(h)}(\mathbf{L}') = \sum_{k=1}^{n} \frac{1}{k}\, T_n^{(h)}(k),$$

where $\mathbf{L}' = (-1, 0, \ldots, 0)$. Now estimates (4.36) follow from Lemma 6. $\square$

Theorem 4.5 *As* $n \to \infty$, *the asymptotic distribution of the number* θ *of images of a fixed element with respect to a random mapping is given by the relations*

$$(4.37) \quad \mathbf{P}\{\theta = \nu\}$$

$$\begin{cases} = \dfrac{1}{r_0 n} \sum_{j=0}^{\nu} r_j \left(1 + o(1)\right), & 1 \le \nu, \\[2mm] & h \ge \nu, \\[2mm] = \dfrac{1}{r_0 n}\left(1 + o(1)\right), & h + 1 \le \nu, \\[2mm] & r_0 n - n^{2/3-\varepsilon} \ge \nu, \\[2mm] \le \dfrac{1}{r_0 n}\left(1 + o(1)\right), & r_0 n - n^{2/3-\varepsilon} + 1 \le \nu, \\[2mm] & r_0 n + n^{2/3-\varepsilon} + h \ge \nu, \\[2mm] = o\left(\dfrac{1}{n}\right), & r_0 n + n^{2/3-\varepsilon} + h + 1 \le \nu, \\[2mm] & n \ge \nu. \end{cases}$$

Proof   From (4.32) we have, for $0 \leq l \leq h$,

$$\mathbf{P}\{v = l, \, w = j\} = \frac{r_j}{r_0 \, n} \left(1 + o(1)\right), \qquad\qquad n \to \infty.$$

This proves the first equality in (4.37). The remaining relations also follow from formula (4.32).   □

6.5   Random mappings with constraints on the height and cyclic components

6.5.1   Asymptotic formulae for mappings

We put $\mathbf{A} = \{1, 2, \ldots, s\}$, $s \geq 1$, and denote by $\mathfrak{G}_n^h(\mathbf{A})$ the set of all mappings of an n-element set $\mathbf{X}$ onto itself whose height belongs to the set $\mathbf{A}$. If $T_n^{(h)}(k, j; s)$ is the number of mappings $\sigma \in \mathfrak{G}_n^h(\mathbf{A})$ with k cyclic elements and j components and

$$(5.1) \qquad Q_n^{(s)}(x, y, h) := \sum_{k=0}^{n} \sum_{j=0}^{k} T_n^{(h)}(k, j; s) \, x^k \, y^j$$

then, by formula (0.15),

$$(5.2) \qquad \exp\left\{y \, A\big(x \, a_h(t)\big)\right\} = \sum_{n=0}^{\infty} Q_n^{(s)}(x, y, h) \, \frac{t^n}{n!} \, ,$$

where

$$A(w) := \sum_{j=1}^{s} \frac{w^j}{j}$$

and

$$(5.3) \qquad a_h(t) = t \exp\left\{t \exp\left\{t \exp\left\{\ldots \exp t\right\}\right\}\right\}$$

(the symbol exp is repeated h times) is the generating function of the number of rooted trees with height not exceeding h.

Theorem 5.1   *Let* $\log_{(1)} z := \log z$, $\log_{(k)} z := \log\left(\log_{(k-1)} z\right)$, $k = 2, 3, \ldots$,

$$(5.4) \qquad \Pi_n(h, s, x, y) := \prod_{j=1}^{h} \log_{(j)} \left(\frac{1}{x}\left(\frac{n}{y}\right)^{1/s}\right),$$

and let $R = R(n, x, y)$ be the unique real solution of the equation

$$(5.5) \qquad A'\left(x\, a_h(R)\right) a_h'(R)\, R = \frac{n}{x\, y}\,,$$

where primes denote differentiation with respect to R. Then there exists a neighborhood $W = [1-x_0 \le x \le 1+x_0,\ 1-y_0 \le y \le 1+y_0]$, $0 < x_0, y_0 < 1$, of the point $(x, y) = (1, 1)$ in which the asymptotic representation

$$(5.6) \qquad Q_n^{(s)}(x, y, h) = \frac{n!\,\exp\left\{y\, A\left(x\, a_h(R)\right)\right\}}{R^n \sqrt{2\,\pi\, s\, n\, \Pi_n(h, s, x, y)}}\left(1 + o(1)\right)$$

is valid as $n \to \infty$ with $o(1) \to 0$ uniformly for $(x, y) \in W$.

Proof Applying Cauchy's formula we find

$$Q_n^{(s)}(x, y, h) = \frac{n!}{2\pi i}\oint_C \exp\left\{y\, A\left(x\, a_h(z)\right)\right\}\frac{dz}{z^{n+1}}\,,$$

where C is any closed contour in the complex plane which encloses the origin. Taking as C the circle of radius $R = R(n, x, y)$ we write

$$Q_n^{(s)}(x, y, h) = \frac{n!}{2\pi R^n}\,\exp\left\{y\, A\!\left(x\, a_h(R)\right)\right\}\times\, J\,,$$

where

$$(5.7) \qquad J := \int_{-\pi}^{\pi} \exp\left\{f(\theta, R, x, y)\right\}d\theta$$

and

$$f(\theta, R, x, y) := y\left\{A\left(x\, a_h\!\left(R e^{i\theta}\right)\right) - A\left(x\, a_h(R)\right)\right\} - i n\theta\,.$$

To prove the theorem it suffices to show that by choosing for R the solution of equation (5.5) we obtain

$$(5.8) \qquad J = \sqrt{\frac{2\,\pi}{s\, n\, \Pi_n(h, s, x, y)}}\left(1 + o(1)\right),$$

where $o(1) \to 0$ uniformly for $(x, y) \in W$.

Take $\varepsilon = n^{-2/5}$ and represent the integral J as follows:

$$J = \int_{-\pi}^{\pi} = \int_{-\pi}^{-\varepsilon} + \int_{-\varepsilon}^{\varepsilon} + \int_{\varepsilon}^{\pi} = J_1 + J_2 + J_3\,.$$

Now the formula will be proved once we show that

$$(5.9) \qquad J_2 = \sqrt{\frac{2\,\pi}{s\,n\,\Pi_n(h,s,x,y)}}\,\left(1+o(1)\right),$$

$$(5.10) \qquad \begin{cases} J_1 = O\!\left(\exp\left\{-\gamma_1\,n^{1/5}\right\}\right), \\ J_3 = O\!\left(\exp\left\{-\gamma_3\,n^{1/5}\right\}\right), \end{cases}$$

where $o(1) \to 0$ uniformly for $(x,y) \in W$ and $\gamma_1 > 0$ and $\gamma_3 > 0$ are independent of x and y for $(x,y) \in W$.

In order to verify relations (5.9) and (5.10) we will prove several auxiliary assertions. Set

$$L_0 = R, \qquad L_k = R \exp\left\{L_{k-1}\right\}, \qquad k = 1, 2, \ldots, h,$$

noting that $L_h(R) = a_h(R)$. As $n \to \infty$, equation (5.5) has a unique real solution $R = R(n,x,y)$, which is a monotonically decreasing function of x and y; moreover, $R \to \infty$ as $n \to \infty$ for $0 < x,\,y < \infty$. Writing equation (5.5) in the form

$$\sum_{l=0}^{h} L_l\,L_{l+1}\cdots L_h \cdot \sum_{j=1}^{s} (xL_h)^{j-1} = \frac{n}{x\,y}\,,$$

we obtain the relation

$$L_0\,L_1 \cdots L_{h-1}\,L_h^s = \frac{n}{x^s\,y}\,\left(1+o(1)\right),$$

where $o(1) \to 0$ uniformly for $(x,y) \in W$. This formula yields the following asymptotic representations:

$$(5.11) \qquad \begin{cases} L_{h-k} = \log_{(k)}\left(\dfrac{1}{x}\left(\dfrac{n}{y}\right)^{1/s}\right) \\ \qquad\quad - \log R + o(1), \quad k = 2, 3, \ldots, h, \\[2mm] L_{h-1} = \log\left(\dfrac{1}{x}\left(\dfrac{n}{y}\right)^{1/s}\right) - \dfrac{1}{s}\,\log_{(2)}\left(\dfrac{1}{x}\left(\dfrac{n}{y}\right)^{1/s}\right) \\ \qquad\quad -\dfrac{1}{s}\sum_{j=0}^{h-2}\log L_j - \log R + o(1), \\[2mm] L_h = \dfrac{1}{x}\left(\dfrac{n}{y}\right)^{1/s}\,\Pi_n^{-1/s}(h,s,x,y)\,\left(1+o(1)\right), \end{cases}$$

where in every case $o(1) \to 0$ uniformly for $(x, y) \in W$. On applying these formulae, asymptotic estimates for the derivatives

$$L_k^{(\nu)} = \frac{d^\nu}{dR^\nu} L_k, \qquad k = 0, 1, \dots,$$

can also be obtained. We have

$$L_1^{(\nu)} = L_1 \left(1 + \frac{\nu}{R} \right),$$

$$L_k^{(\nu)} = (L_1 L_2 \cdots L_{k-1})^\nu L_k (1 + o(1)), \qquad k = 1, 2, \dots.$$

The proof of these formulae is based on the relation

$$L_k^{(\nu)} = R \left(\exp\{L_{k-1}\} \right)^{(\nu)} + \nu \left(\exp\{L_{k-1}\} \right)^{(\nu-1)},$$

which can be obtained by Leibnitz's formula applied when differentiating the equation $L_k = R \exp\{L_{k-1}\}$, and on the asymptotic representation

$$\left(\exp\{L_{k-1}\} \right)^{(\nu)} = (L_1 L_2 \cdots L_{k-1})^\nu \exp\{L_{k-1}\} (1 + o(1)),$$

which can be deduced by appeal to Bruno's formula for differentiation of composite functions and by mathematical induction on k:

$$\left(\exp\{L_{k-1}\} \right)^{(\nu)} = \sum_{\Sigma j \, \alpha_j = \nu} \frac{\nu!}{\alpha_1! \cdots \alpha_\nu!} \exp\{L_{k-1}\}$$

$$\times \left(\frac{1}{1!} L_{k-1}^{(1)} \right)^{\alpha_1} \cdots \left(\frac{1}{\nu!} L_{k-1}^{(\nu)} \right)^{\alpha_\nu}$$

$$= (L_1 L_2 \cdots L_{k-2})^\nu \exp\{L_{k-1}\} \sum_{\alpha=1}^{\nu} \sigma(\nu, \alpha) L_{k-1}^\alpha (1 + o(1))$$

$$= (L_1 L_2 \cdots L_{k-1})^\nu \exp\{L_{k-1}\} (1 + o(1)),$$

where $\sigma(\nu, \alpha)$ are Stirling's numbers of the second kind and in all the formulae $o(1) \to 0$ uniformly for $(x, y) \in W$.

Using these results it is not difficult to derive the formulae

(5.12) $a_{1h}(R) = R a_h'(R) = L_0 L_1 \cdots L_h (1 + o(1)),$

(5.13) $a_{2h}(R) = R a_{1h}'(R)$

$$= \frac{a_{1h}^2(R)}{a_h(R)} + (L_0 L_1 \cdots L_{h-2})^2 L_{h-1} L_h (1 + o(1)),$$

where primes signify differentiation and $o(1) \to 0$ uniformly for $(x, y) \in W$. Indeed, if

$$M_k = R L_k^{(1)}, \qquad\qquad k = 0, 1, \ldots, h,$$

then

$$M_k = L_k \left(M_{k-1} + 1 \right), \qquad k = 1, 2, \ldots, h,$$

that is,

$$M_k = \sum_{j=0}^{k} L_j L_{j+1} \cdots L_k, \qquad k = 0, 1, \ldots, h.$$

Hence formula (5.12) follows. Relation (5.13) can be established from the equation

$$a_{2h}(R) = \frac{a_{1h}^2(R)}{a_h(R)} + \sum_{j=0}^{h-1} \frac{L_{j+1} L_{j+2} \cdots L_h}{L_j} \left(\sum_{\nu=0}^{j} L_\nu L_{\nu+1} \cdots L_j \right)^2.$$

The results obtained allow us to prove the following statement:

Lemma 1   *Uniformly for all* θ, $-\varepsilon \le \theta \le \varepsilon$,

$$f(\theta, R, x, y) = \tfrac{1}{2} f''(0, R, x, y)\, \theta^2 + o(1),$$

$$f''(0, R, x, y) = -s\, n\, \Pi_n(h, s, x, y)\, (1 + o(1))$$

as $n \to \infty$, *where* $o(1) \to 0$ *uniformly for* $(x, y) \in W$. *There is an absolute positive constant* γ *such that, for* $\varepsilon \le \theta \le \pi$,

$$\mathrm{Re}\, f(\theta, R, x, y) \le -\gamma\, n^{1/5}, \qquad n \to \infty.$$

Proof   We expand $f(\theta, R, x, y)$ in the neighborhood $-\varepsilon \le \theta \le \varepsilon$ in a Maclaurin series:

$$f(\theta, R, x, y) = \tfrac{1}{2} f''(0, R, x, y)\, \theta^2 + H(\theta, R, x, y),$$

$$H(\theta, R, x, y) = \sum_{\nu=3}^{\infty} f^{(\nu)}(0, R, x, y)\, \frac{\theta^\nu}{\nu!}.$$

Applying Bruno's formula once again to finding the νth derivative of f gives

$$f^{(\nu)}(0, R, x, y) = i^\nu y\, x^s\, s^{\nu-1} \left(L_0 L_1 \cdots L_{h-1} \right)^\nu L_h^s\, (1 + o(1))$$

$$= i^\nu s^{\nu-1} n\, \Pi_n^{\nu-1}(h, s, x, y)\, (1 + o(1)),$$

where $o(1) \to 0$ uniformly for $(x,y) \in W$. Hence we conclude that

$$\left| H(\theta, R, x, y) \right| \le n^{-1/5} \left[s\, \Pi_n(h, s, x, y) \right]^2 \exp\left\{ s\, n^{-2/5}\, \Pi_n(h, s, x, y) \right\}.$$

The first assertion of the lemma immediately follows from this relation.
Further, from the equation $\operatorname{Re} L_0(R e^{i\theta}) = R \cos\theta$ and the estimates

$$\operatorname{Re} L_k(R e^{i\theta}) \le R \exp\left\{ \operatorname{Re} L_{k-1}(R e^{i\theta}) \right\},$$

$$\operatorname{Re} L_k^j(R e^{i\theta}) \le R^j \exp\left\{ j\, [\operatorname{Re} L_{k-1}(R e^{i\theta})] \right\},$$

setting

$$a_1(t,y) := t \exp\{f\, y\}, \quad a_k(t, y_1, \dots, y_k) := t \exp\left\{ y_1 a_{k-1}(t, y_2, \dots, y_k) \right\},$$

we conclude that

$$\operatorname{Re} L_k^j(R e^{i\theta}) \le a_k^j(R, 1, \dots, 1, \cos\theta).$$

Thus, for $\varepsilon \le \theta \le \pi$

$$\operatorname{Re} L_k^j(R e^{i\theta}) - L_k^j(R) \le a_k^j(R, 1, \dots, 1, \cos\varepsilon) - a_k^j(R, 1, \dots, 1, 1).$$

Hence we obtain by induction

$$\operatorname{Re} L_k^j(R e^{i\theta}) - L_k^j(R) \le -\tfrac{1}{2} j\, L_0 L_1 \cdots L_{k-1} L_k^j \varepsilon^2 + o(1),$$

$$k = 1, 2, \dots, h, \qquad j = 1, 2, \dots, s.$$

From these inequalities the following estimate, valid for all $(x,y) \in W$, can be deduced:

$$\operatorname{Re} L_h^s(R e^{i\theta}) - L_h^s(R) \le -\gamma\, n^{1/5},$$

where γ is an absolute positive constant. This implies the validity of the second part of the lemma. $\quad\square$

We now note that, by Lemma 1,

$$J_2 = \int_{-\varepsilon}^{\varepsilon} \exp\left\{ \tfrac{1}{2} f''(0, R, x, y)\theta^2 \right\} d\theta\, (1 + o(1)),$$

where $o(1) \to 0$ uniformly for $(x,y) \in W$. Hence,

$$J_2 = \frac{1}{\sqrt{-f''(0, R, x, y)}} \int_{-\delta}^{\delta} e^{-\varphi^2/2}\, d\varphi\, (1 + o(1)),$$

$$\delta = \sqrt{s}\, n^{1/10}\, \Pi_n^{1/2}(h, s, x, y)\, (1 + o(1)),$$

where $o(1) \to 0$ uniformly for $(x,y) \in W$. The validity of formula (5.9) follows from these equations, while the second part of Lemma 1 assures us of the validity of estimates (5.10). Theorem 5.1 is proved. $\quad\square$

Corollary 1   *Let s be a fixed number. Then the number of mappings in $\mathfrak{G}_n^h(\mathbf{A})$, $\mathbf{A} = \{1, 2, \dots, s\}$, has the asymptotic representation*

$$(5.14) \qquad T_n^{(h)}(\mathbf{A}) = \left(\frac{n}{r}\right)^n \exp\left\{A\left(a_h(r)\right) - n\right\}$$

$$\times \left(s \prod_{j=1}^{h} \log_{(j)} n^{1/s}\right)^{-1/2} \left(1 + o(1)\right)$$

as $n \to \infty$, where r is, for large n, the unique real solution of the equation

$$A'\left(a_h(r)\right) a'_h(r) r = n.$$

An asymptotic representation, with explicit dependence on n only, is of the form

$$\frac{1}{n} \log T_n^{(h)}(\mathbf{A}) = \log n - \log_{(h+1)} n^{1/s} - 1 + \left(s \prod_{j=1}^{h} \log_{(j)} n^{1/s}\right)^{-1} + o(1),$$

where $\prod_{j=1}^{0} \log_{(j)} n^{1/s} = 1$.

Formula (5.14) follows from formula (5.6) if we set $x = y = 1$.

Corollary 2   *The number of labeled rooted forests with n vertices and with heights of trees not exceeding h or, what amounts to the same, the number of elements σ of the symmetric semigroup $\mathfrak{G}_n$ satisfying the condition $\sigma^{h+1} = \sigma^h$ is given by the asymptotic formula*

$$(5.15) \qquad T_n^{(h)} = \left(\frac{n}{r}\right)^n \exp\left\{a_h(r) - n\right\} \left(\prod_{j=1}^{h} \log_{(j)} n\right)^{-1/2} \left(1 + o(1)\right)$$

as $n \to \infty$, where r is the unique real solution of the equation $r\, a'_h(r) = n$.

Relation (5.15) can be obtained from (5.14) by merely setting $s = 1$.

Finally, we formulate a result by Harris and Schoenfeld [55] concerning the asymptotic behavior of I_n, the number of idempotents in the symmetric semigroup $\mathfrak{G}_n$, that is, the number of elements $\sigma \in \mathfrak{G}_n$ satisfying the condition $\sigma^2 = \sigma$ or, what amounts to the same, the number of labeled forests consisting of trees whose height is either 1 or 0.

Corollary 3   *If r is the unique real root of the equation $r(r+1)e^r = n$ then, as $n \to \infty$, the following asymptotic representation is valid:*

$$(5.16) \qquad I_n = \frac{1}{\sqrt{\log n}} \exp\left\{n\left(\log n - \log r - 1 + \frac{1}{r+1}\right)\right\} \left(1 + o(1)\right).$$

Corollary 3 is a particular case of Corollary 2 for $h = 1$.

6.5.2 Asymptotic formulae for the number of trees
of a bounded height

Using generating function (5.3) we can represent the number $D_n^{(h)}$ of rooted trees of height not exceeding h as

$$D_n^{(h)} = \frac{n!}{2\pi i} \oint_C a_h(z) \frac{dz}{z^{n+1}},$$

where C is a closed contour in the complex plane z enclosing the origin. Taking as C the circle of radius $\tilde{r}$ with center at the origin we obtain

$$D_n^{(h)} = \frac{n!\, a_h(\tilde{r})}{2\,\pi\,\tilde{r}^n} \times J, \qquad J = \int_{-\pi}^{\pi} \exp\{f(\theta,\tilde{r})\}\, d\theta,$$

$$f(\theta,\tilde{r}) = L_{h-1}(\tilde{r}\, e^{i\theta}) - L_{h-1}(\tilde{r}) - i\,\theta\,(n-1).$$

Setting

$$a_{hk}(x) := \left(x\, \frac{d}{dx}\right)^k L_{h-1}(x)$$

and choosing for $\tilde{r}$ a unique (as $n \to \infty$) real solution of the equation

(5.17) $$a_{h1}(\tilde{r}) = n - 1,$$

one can show, applying the procedure used in Theorem 1, that

$$J = \sqrt{\frac{2\,\pi}{a_{h2}(\tilde{r})}}\, \left(1 + o(1)\right).$$

Therefore,

(5.18) $$D_n^{(h)} = \frac{n!\, a_h(\tilde{r})}{\tilde{r}^n \sqrt{2\,\pi\, a_{h2}(\tilde{r})}}\, \left(1 + o(1)\right).$$

Equation (5.17) can be written in the form

(5.19) $$\sum_{j=0}^{h-1} L_j(\tilde{r})\, L_{j+1}(\tilde{r})\, \cdots\, L_{h-1}(\tilde{r}) = n - 1.$$

Hence one can deduce the formulae

(5.20) $$L_{h-j-1}(\tilde{r}) = \log_{(j)} n\, \left(1 + o(1)\right), \qquad j = 1, 2, \ldots, h - 1.$$

We note, in particular, that

$$(5.21) \qquad \tilde{r} = L_0(\tilde{r}) = \log_{(h-1)} n \left(1 + o(1)\right).$$

Furthermore, equalities (5.19) and (5.20) imply the asymptotic representation

$$(5.22) \qquad L_{h-1}(\tilde{r}) = n \left(\prod_{j=1}^{h-1} \log_{(j)} n \right)^{-1} \left(1 + o(1)\right),$$

where we agree to consider $\prod_{j=1}^{0} \log_{(j)} n = 1$.

By using the asymptotic estimates it is not difficult to deduce the following formulae

$$(5.23) \qquad a_h(\tilde{r}) = \log_{(h-1)} n \, \exp \left\{ n \left(\prod_{j=1}^{h-1} \log_{(j)} n \right)^{-1} \left(1 + o(1)\right) \right\},$$

$$(5.24) \qquad a_{h2}(\tilde{r}) = n \prod_{j=1}^{h-1} \log_{(j)} n \left(1 + o(1)\right).$$

Applying Stirling's formula and formula (5.24) we see from (5.18) that the following theorem is valid.

Theorem 5.2 *For any fixed h and $n \to \infty$, the number of labeled rooted trees with n vertices and height not exceeding h is asymptotically equal to*

$$(5.25) \qquad D_n^{(h)} = \left(\frac{n}{e\tilde{r}} \right)^n a_h(\tilde{r}) \left(\prod_{j=1}^{h-1} \log_{(j)} n \right)^{-1/2} \left(1 + o(1)\right),$$

where $a_h(t) = t \exp\{t \exp\{t \exp\{\ldots \exp t\}\}\}$ is the same as in (5.3) and r is, for large n, the unique real solution of equation (5.17).

Hence the following asymptotic representation with dependence on n only can be deduced:

$$(5.26) \qquad \frac{1}{n} \log D_n^{(h)} = \log n - \log_{(h)} n - 1 + \left(\prod_{j=1}^{h-1} \log_{(j)} n \right)^{-1} + o(1).$$

Let us consider some particular cases of the formulae obtained. Setting $h = 1$ we see that $\tilde{r} = n - 1$ and $a_1(\tilde{r}) = (n-1) e^{n-1}$ and, therefore,

$$D_n^{(1)} = n \left(1 + o(1)\right).$$

This fact is obvious since the number of rooted trees of height 1 with n vertices is equal to n.

We now consider the case $h = 2$. If I_n is the number of forests with n vertices constituted by trees of height not exceeding 1 then $D_n^{(2)} = n\,I_{n-1}$. From (5.16) we derive

$$I_{n-1} = \frac{1}{n\sqrt{\log n}}\left(\frac{n}{er}\right)^n r \exp\left\{r\,e^r\right\}(1+o(1)),$$

where

$$r\,(r+1)\,e^r = n - 1.$$

Hence the asymptotic relation

$$(5.27)\qquad D_n^{(2)} = \frac{1}{\sqrt{\log n}}\left(\frac{n}{e\tilde{r}}\right)^n \tilde{r}\,\exp\left\{\tilde{r}\,e^{\tilde{r}}\right\}(1+o(1)),$$

$$(5.28)\qquad \tilde{r}\,(\tilde{r}+1)\,e^{\tilde{r}} = n - 1$$

follows, which is a particular case of (5.25) since equation (5.28) is equation (5.17) for $h = 2$.

6.5.3 A-trees and A-permutations

Let $\mathbf{A}$ be a subset of nonnegative integers. An unrooted forest is called an $\mathbf{A}$-*forest* if the number of vertices of any tree of the forest belongs to the set $\mathbf{A}$. For an $\mathbf{A}$-permutation of order n having k cycles and belonging to a cycle class $[1^{\alpha_1} 2^{\alpha_2} \cdots n^{\alpha_n}]$ we denote by $U(\alpha_1,\alpha_2,\dots,\alpha_n;\mathbf{A})$ the number of representations of the permutation in the form of a composition of $n-k$ transpositions (we note that any representation of the $\mathbf{A}$-permutation in the form of a composition of transpositions consists of at least $n-k$ transpositions). The following formula is valid:

$$(5.29)\qquad U(\alpha_1,\alpha_2,\dots,\alpha_n;\mathbf{A}) = (n-k)!\prod_{j\in\mathbf{A}}\left(\frac{j^{j-2}}{(j-1)!}\right)^{\alpha_j},$$

where, according to the conditions given above,

$$\sum_{j\in\mathbf{A}} j\,\alpha_j = n,\qquad \sum_{j\in\mathbf{A}} \alpha_j = k.$$

From formula (5.29) we obtain an expression for the quantity $U_k(n;\mathbf{A})$, the number of products of $n-k$ transpositions generating $\mathbf{A}$-permutations:

$$(5.30)\qquad U_k(n;\mathbf{A}) = n!\,(n-k)!\sum_{\substack{\sum j\alpha_j=n\\ \sum \alpha_j=k\\ j\in\mathbf{A}}}\prod_{j\in\mathbf{A}}\frac{1}{\alpha_j!}\left(\frac{j^{j-2}}{j!}\right)^{\alpha_j}.$$

On the other hand, if $\tilde{r}_n^{(k)}(\mathbf{A})$ is the number of $\mathbf{A}$-forests with n vertices and k trees then

$$(5.31) \qquad \tilde{r}_n^{(k)}(\mathbf{A}) = n! \sum_{\substack{\Sigma j \alpha_j = n \\ \Sigma \alpha_j = k \\ j \in \mathbf{A}}} \prod_{j \in \mathbf{A}} \frac{1}{\alpha_j!} \left(\frac{j^{j-2}}{j!} \right)^{\alpha_j}.$$

Formulae (5.30) and (5.31) imply the relation

$$(5.32) \qquad U_k(n; \mathbf{A}) = \tilde{r}_n^{(k)}(\mathbf{A}) \, (n-k)!.$$

We now consider the generating function for $\mathbf{A}$-forests:

$$D(t, x; \mathbf{A}) = \sum_{n=0}^{\infty} \sum_{k=0}^{n} \tilde{r}_n^{(k)}(\mathbf{A}) \, \frac{t^n}{n!} \, x^k, \qquad \tilde{r}_0^{(0)}(\mathbf{A}) = 1.$$

It follows from the foregoing that

$$(5.33) \qquad D(t, x; \mathbf{A}) = \exp \left\{ x \, E(t; \mathbf{A}) \right\},$$

where

$$E(t; \mathbf{A}) := \sum_{j \in \mathbf{A}} j^{j-2} \, \frac{t^j}{j!}.$$

Theorem 5.3 *Let a set* $\mathbf{A} = \{j_1, j_2, \ldots, j_m\}$ *be fixed, where* $m > 1$, $j_1 < j_2 < \cdots < j_m$ *and* $\gcd(j_1, j_2, \ldots, j_m) = 1$. *As* $n \to \infty$, *the number of unlabeled* $\mathbf{A}$-forests with n vertices is given by the asymptotic formula

$$(5.34) \qquad \tilde{r}_n^{(k)}(\mathbf{A}) = j_m^{-1/2} \left(\frac{n}{er} \right)^n \exp \left\{ E(r; \mathbf{A}) \right\} \left(1 + o(1) \right),$$

where, for large n, r *is the maximal positive root of the equation*

$$(5.35) \qquad \sum_{k=1}^{m} j_k^{j_k - 1} \frac{r^{j_k}}{j_k!} = n.$$

Proof We make use of the corollary to Theorem 4.1 of Chapter 1. We have

$$g(t) = E(t; \mathbf{A}) = \sum_{k=1}^{m} j_k^{j_k - 2} \frac{t^{j_k}}{j_k!}.$$

The equation determining r is as follows:

$$r \, E'(t; \mathbf{A}) = n,$$

that is, it coincides with equation (5.35). Thus,

$$r = \left(\frac{n \, j_m!}{j_m^{j_m - 1}} \right)^{1/j_m} \left(1 + o(1) \right).$$

Taking into account the relation

$$\left(r \, \frac{d}{dr} \right)^2 E(r; \mathbf{A}) = n \, j_m \, (1 + o(1))$$

we arrive at (5.34). $\qquad \square$

6.5.4 Distribution of the number of cyclic elements
and components in a random A-mapping

We consider the uniform probability distribution on the set $\mathfrak{G}_n^h(\mathbf{A})$, $\mathbf{A} = \{1, 2, \ldots, s\}$, and denote by $\xi_n(s, h)$ and $\varkappa_n(s, h)$ the numbers of cyclic elements and components, respectively, in a random mapping $\sigma \in \mathfrak{G}_n^h(\mathbf{A})$. According to the results of Subsection 6.5.1, the generating functions of $\xi_n(s, h)$ and $\varkappa_n(s, h)$ have the respective forms:

$$P_n(x; s, h) = \frac{Q_n^{(s)}(x, 1, h)}{Q_n^{(s)}(1, 1, h)}, \qquad \widetilde{P}_n(y; s, h) = \frac{Q_n^{(s)}(1, y, h)}{Q_n^{(s)}(1, 1, h)}.$$

Below we will use the notations and results of Theorem 5.1. To this aim we denote by $R_1 = R(n, x, 1)$ and $R_2 = R(n, 1, y)$ the unique real solutions of the equations

$$(5.36) \qquad \begin{cases} A'\big(xa_h(R_1)\big)\, a_h'(R_1)\, R_1 = n/x\,, \\ A'\big(a_h(R_2)\big)\, a_h'(R_2)\, R_2 = n/y \end{cases}$$

respectively. Further, let

$$r = R(n, 1, 1)\,, \qquad R_1 = R(n, x, 1)\,, \qquad R_2 = R(n, 1, y)$$

and

$$(5.37) \qquad \begin{cases} g_n(x) = A\big(xa_h(R_1)\big) - A\big(a_h(r)\big) \\ \qquad +n \log \dfrac{r}{R_1} + \dfrac{1}{2} \log \dfrac{\Pi_n(h, s, 1, 1)}{\Pi_n(h, s, x, 1)}\,, \\ v_n(y) = y\, A\big(a_h(R_2)\big) - A\big(a_h(r)\big) \\ \qquad +n \log \dfrac{r}{R_2} + \dfrac{1}{2} \log \dfrac{\Pi_n(h, s, 1, 1)}{\Pi_n(h, s, 1, y)}\,. \end{cases}$$

From Theorem 5.1 it follows that, as $n \to \infty$,

$$(5.38) \qquad \begin{cases} P_n(x; s, h) = \exp\big\{g_n(x)\big\}\,\big(1 + o(1)\big)\,, \\ \widetilde{P}_n(y; s, h) = \exp\big\{v_n(y)\big\}\,\big(1 + o(1)\big)\,, \end{cases}$$

where in both relations $o(1) \to 0$ uniformly for $x \in W_1 = [\,1 - x_0, 1 + x_0\,]$, $0 < x_0 < 1$, and $y \in W_2 = [\,1 - y_0, 1 + y_0\,]$, $0 < y_0 < 1$, respectively.

It is convenient to introduce the notations:

$$M_1 := A'\big(a_h(r)\big)\,a_h(r)\,, \qquad\qquad M_2 := A\big(a_h(r)\big)\,,$$

$$\sigma_1^2 := \frac{A''\big(a_h(r)\big)\,A'\big(a_h(r)\big)\,a_h(r) + \big\{A'\big(a_h(r)\big)\big\}^2}{A''\big(a_h(r)\big)\,a_{1h}^2(r) + A'\big(a_h(r)\big)\,a_{2h}(r)}\ \big(a_h(r)\,a_{2h}(r) - a_{1h}^2(r)\big)\,,$$

$$\sigma_2^2 := A\big(a_h(r)\big) - \frac{\big\{A'\big(a_h(r)\big)\,a_{1h}(r)\big\}^2}{A''\big(a_h(r)\big)\,a_{1h}^2(r) + A'\big(a_h(r)\big)\,a_{2h}(r)}\,,$$

where primes denote differentiation and the quantities a_{1h} and a_{2h} are determined by relations (5.12) and (5.13) respectively.

We now formulate two theorems describing the limiting behavior of the distributions of $\xi_n(s,h)$ and $\varkappa_n(s,h)$.

Theorem 5.4 *For any fixed s and h, the distribution of the random variable $\eta_1 = (\,\xi_n(s,h) - M_1\,)\,\sigma_1^{-1}$ converges to the standard normal distribution as $n \to \infty$.*

Theorem 5.5 *For any fixed s and h, the distribution of the random variable $\eta_2 = (\,\varkappa_n(s,h) - M_2\,)\,\sigma_2^{-1}$ converges to the standard normal distribution as $n \to \infty$.*

The proofs of the theorems are similar and can be carried out in parallel. In preparation for this, put

$$L_j := L_j(R_1)\,, \qquad \bar{L}_j := L_j(R_2)\,, \qquad j = 1, 2\ldots, h - 1\,,$$

$$R_1^{(\mu)} := \frac{d^\mu}{dx^\mu}\,R_1\,, \qquad\qquad R_2^{(\mu)} := \frac{d^\mu}{dx^\mu}\,R_2\,.$$

Lemma 2 *For any fixed s and h, the following asymptotic relations are valid as $n \to \infty$:*

$$R_1^{(\mu)} = \frac{(-1)^\mu\,(\mu - 1)!}{x^\mu\,L_1 L_2 \ldots L_{h-1}}\,\big(1 + o(1)\big)\,,$$

$$R_2^{(\mu)} = \frac{(-1)^\mu\,(\mu - 1)!}{s\,y^\mu\,\bar{L}_1 \bar{L}_2 \ldots \bar{L}_{h-1}}\,\big(1 + o(1)\big)\,,$$

where, in both formulae, $o(1) \to 0$ uniformly for $x \in W_1$ and $y \in W_2$ respectively.

The proof of Lemma 2 consists of taking the μth derivatives of both sides of equations (5.36).

Lemma 3 *If $g_n^{(k)}(x)$ and $v_n^{(k)}(y)$ are the kth derivatives of the functions $g_n(x)$ and $v_n(y)$, respectively, then, for any fixed s and h, the following relations are valid as $n \to \infty$:*

$$g_n^{(k+1)}(x) = (-1)^k \, x^{s-k-1} \, k! \, L_h^s(R_1) \, (1 + o(1)) = O\big(n \, \Pi_n^{-1}(h, s, 1, 1) \big),$$

$$v_n^{(k+1)}(y) = \frac{(-1)^k \, k!}{s \, y^k} \, L_h^s(R_2) \, (1 + o(1)) = O\big(n \, \Pi_n^{-1}(h, s, 1, 1) \big),$$

where all the estimates are uniform for $x \in W_1$ and $y \in W_2$ respectively.

The proof of Lemma 3 is straightforward and is based on the results of Lemma 2. From Lemma 3, relations (5.11)–(5.13) and (5.37) we conclude that, as $n \to \infty$,

$$M_1 = g_n'(1) = \frac{n}{\Pi_n(h, s, 1, 1)} \, (1 + o(1)),$$

$$M_2 = v_n'(1) = \frac{n}{s \, \Pi_n(h, s, 1, 1)} \, (1 + o(1)),$$

$$\sigma_1^2 = g_n'(1) + g_n''(1) = \frac{s \, n}{\log n \, \Pi_n(h, s, 1, 1)} \, (1 + o(1)),$$

$$\sigma_2^2 = v_n'(1) + v_n''(1) = \frac{n}{s \log n \, \Pi_n(h, s, 1, 1)} \, (1 + o(1)).$$

Moreover, uniformly for $x \in W_1$ and $y \in W_2$, respectively, we have

$$g_n'''(x) \, \sigma_1^{-3} \to 0, \qquad v_n'''(y) \, \sigma_2^{-3} \to 0,$$

$$\lim_{n \to \infty} \frac{\Pi_n(h, s, 1, 1)}{\Pi_n(h, s, x, 1)} = \lim_{n \to \infty} \frac{\Pi_n(h, s, 1, 1)}{\Pi_n(h, s, 1, y)} = 1.$$

Taking representations (5.38) into account, we can now apply Theorem 4.2 of Chapter 1. This completes the proof of Theorems 5.4 and 5.5. $\square$

Let F_n^h be a labeled rooted forest with n vertices whose trees are of height not exceeding h, and let $\mathcal{F}_n^h$ be the family of all such forests.

Corollary 1 *If ξ_n is the number of trees in a labeled rooted forest chosen at random and equiprobably from $\mathcal{F}_n^h$ then the distribution of the random variable*

$$\big(\xi_h - a_h(r) \big) \big(a_h(r) - a_{1h}^2(r)/a_{2h}(r) \big)^{-1/2},$$

where r is the unique real solution of the equation $a_{1h}(r) = n$, converges to the standard normal distribution as $n \to \infty$.

Corollary 2 *If ξ_1 is the number of fixed elements of an idempotent chosen at random and equiprobably from the symmetric semigroup $\mathfrak{S}_n$ then the distribution of the random variable*

$$\left(\xi_1 - \frac{n}{r+1} \right) \left(\frac{n \, r}{(r+1)^3} \right)^{-1/2}, \qquad r \, (r+1) \, e^r = n,$$

converges to the standard normal distribution as $n \to \infty$.

References

[1] Aleksandrov, A.D. (1938). By the theory of geometric volumes of convex bodies, IV, *Matem. Sbornik*, **3**, No. 2, 227–52 (in Russian).

[2] Austin, T.L., Fagen, R.E., Penney, W.F. and Riordan, J. (1959). The number of components in random linear graphs, *Math. Statist.*, **30**, No. 3, 747–54.

[3] Barton, D.E. and David, F.N. (1962). *Combinatorial Chance*, Griffin, London.

[4] Bekessy, A. (1963–4). On classical occupancy problems, I; II, *Magyar Tud. Acad. Mat. kutato int. Közl.*, **8**, Nos. 1–2; **9** (1964), Nos. 1–2.

[5] Bender, E.A. (1973). Central and local limit theorems applied to asymptotic enumeration, *J. Comb. Theory*, **15**, No. 1, 91–111.

[6] Bender, E.A. and Richmond, L.B. (1983). Central and local limit theorems applied to asymptotic enumeration, II: Multivariate generating functions, *J. Comb. Theory, Ser. A*, **34**, 255–65.

[7] Berge, C. (1958). *Théorie des Graphes et ses Applications*, Dunod, Paris.

[8] Bernstein, S.N. (1946). *The Theory of Probability*, 4th edn, Gostekhizdat, Moscow–Leningrad (in Russian).

[9] Bollobás, B. (1981). Degree sequences of random graphs, *Discr. Math.*, **33**, 1–19.

[10] Bollobás, B. (1984). The evolution of random graphs, *Trans. Amer. Math. Soc.*, **286**, 257–74.

[11] Bollobás, B. (1983). The evolution of sparse graphs, In: *Graph Theory and Combinatorics* (Cambridge, 1983), Academic Press, London–New York, 1984, pp. 35–57.

[12] Bollobás, B. (1985). *Random Graphs*, Academic Press, London.

[13] Bollobás, B. and Thomason, A.G. (1983). Random graphs of small order, In: *Random Graphs* (Poznań, 1983), North-Holland Math. Stud., **118**, North-Holland, Amsterdam–New York, pp. 47–97.

[14] Bolotnikov, Yu.V., Tarakanov, V.E. and Sachkov, V.N. (1976). Asymptotic normality of some variables associated with the cyclic structure of random permutations, *Matem. Sbornik, New Series*, **99**, No. 1, 121–33 (in Russian).

[15] Borovkov, A.A. (1976). *Wahrscheinlichkeitstheorie*, Birkhäuser, Basel–Boston.

[16] de Bruijn, N.G. (1961). *Asymptotic Methods in Analysis*, North-Holland, Amsterdam.

[17] Carlitz, L., Roselle, D.P. and Scoville, R.A. (1966). Permutations and sequences with repetition by number of increases, *J. Comb. Theory*, **1**, No. 3, 350–74.

[18] Cayley, A. (1989–97). *The Collected Mathematical Papers*, Cambridge University Press.

[19] Clarke, L.E. (1958). On Cayley's formula for counting trees, *J. London Math. Soc.*, **33**, No. 132, 471–5.

[20] Cramér, H. (1974). *Mathematical Methods of Statistics*, Princeton University Press.

[21] Curtiss, I.H. (1942). A note on the theory of moment generating functions, *Ann. Math. Statist.*, **13**, No. 3, 430–3.

[22] Dénes, J. (1966). Connections between transformation semigroups and graphs in theory of graphs, In: *Theory of Graphs* (International Symposium, Rome, 1966), Gordon and Breach, New York, 1967, pp. 93–101.

[23] Dénes, J. (1968). On some properties of commutator subsemigroups, *Publ. Math. Debrecen*, **15**, 283–5.

[24] Dénes, J. (1966). On transformations, transformation–semigroups and groups in theory of graphs, In: *Theory of Graphs* (Proceedings of the Colloquium, Tihany, Hungary, 1966), Acad. Kiadó, Budapest, 1968, pp. 65–75.

[25] Dénes, J. (1970). On graph representation of semigroups, In: *Combinatorial Structures and Their Applications* (Proceedings of the Calgary International Conference of Combinatorial Structures and Their Applications), Gordon and Breach, New York, pp. 55–7.

[26] Dénes, J. (1970). Some combinatorial properties of transformations and their connection with the theory of graphs, *J. Comb. Theory*, **9**, No. 2, 108–16.

[27] Dénes, J. and Török, E. (1970). Groups and graphs, In: *Combinatorial Theory and Its Applications*, North-Holland, pp. 257–9.

[28] Dóbinski, V. (1977). Summirung der Reihe $\sum_{k=0}^{\infty} k^m/k!$ für $m = 1, 2, 3, 4, 5, \ldots$, *Grunert Archiv*, **61**, 333–6.

[29] Egorychev, G.P. (1980). *Solution of the van der Waerden Problem for Permanents*, Inst. Fiz. im. L. V. Kirenskogo Sibir. Otd. AN SSSR, Krasnoyarsk IFSO–13M (in Russian).

[30] Érdös, P. and Rényi, A. (1959). On random graphs, I, *Publ. Math. Debrecen*, **6**, 290–7.

[31] Érdös, P. and Rényi, A. (1960). On evolution of random graphs, *Magyar Tud. Acad. Mat. kutato int. Közl.*, **5**, Nos. 1–2, 17–61.

[32] Érdös, P. and Rényi, A. (1961). On the strength of connectedness of a random graph, *Acta Math. Acad. Sci. Hungar.*, **12**, Nos. 1–2, 261–7.

[33] Érdös, P. and Rényi, A. (1961). On a classical problem of probability theory, *Magyar Tud. Acad. Mat. kutato int. Közl.*, **6**, Nos. 1–2, 215–20.

[34] Érdös, P. and Rényi, A. (1963). On random matrices, *Magyar Tud. Acad. Mat. kutato int. Közl.*, **8**, No. 3, 455-61.

[35] Érdös, P. and Rényi, A. (1966). On existence of a factor of degree one of a connected random graph, *Acta Math. Acad. Sci. Hungar.*, **17**, Nos. 3-4, 359-68.

[36] Érdös, P. and Rényi, A. (1968). On random matrices, II, *Studia Sci. Math. Hungar.*, **3**, No. 4, 459-64.

[37] Érdös, P. and Spencer, J. (1974). *Probabilistic Methods in Combinatorics*, Akad. Kiadó, Budapest.

[38] Érdös, P. and Turán, P. (1965). On some problems of a statistical group theory, I, *Z. Wahrsch. Verw. Geb.*, **4**, No. 2, 175-86.

[39] Érdös, P. and Turán, P. (1967). On some problems of a statistical group theory, II; III; IV, *Acta Math Acad. Sci. Hungar.*, **18** (1967), Nos. 1-2, 151-63; **18** (1967), Nos. 3-4, 309-20; **19** (1968), Nos. 3-4, 413-35.

[40] Everett, C.I. and Stein, P.R. (1973). The asymptotic number of $(0,1)$-matrices with zero permanent, *Discr. Math.*, **6**, No. 1, 29-34.

[41] Falikman, D.I. (1981). The proof of the van der Waerden conjecture on the permanent of the double stochastic matrix, *Matem. Zametki*, **29**, 931-8 (in Russian).

[42] Feller, W. (1968-71). *An Introduction to Probability Theory and Its Applications*, Vol. 1, 3rd edn; Vol. 2, 2nd edn, Wiley, New York.

[43] Folkert, I. (1955). *The Distribution of the Number of Components in Random Mapping Functions*, Michigan State University.

[44] Frieze, A.M. (1988). Partitioning random graphs into large cycles, *Discr. Math.*, **70**, 149-58.

[45] Gilbert, E. (1942). Random graphs, *Ann. Math. Statist.*, **13**, 430-3.

[46] Gilbert, E. (1959). Random graphs, *Ann. Math. Statist.*, **30**, 1141-4.

[47] Gnedenko, B.V. (1984). *The Theory of Probability*, 4th edn, Chelsea Publishing Company, New York.

[48] Golomb, S.W. (1964). Random permutations, *Bull. Amer. Math. Soc.*, **70**, No. 6, 747.

[49] Goncharov, V.L. (1944). Some facts from combinatorics, *Izv. AN SSSR, Ser. Matem.*, **8**, No. 1, 3-48 (in Russian).

[50] Harary, F. and Mowshowitz, A. (1969). Labeled trees with unlabeled end-points, *J. Comb. Theory*, **6**, No. 1, 60-4.

[51] Harary, F. and Palmer, E. (1973). *Graphical Anumerition*, Academic Press, New York.

[52] Harper, L.H. (1967). Stirling behavior is asymptotically normal, *Ann. Math. Statist.*, **38**, No. 2, 410-14.

[53] Harris, B. (1960). Probability distributions related to random mappings, *Ann. Math. Statist.*, **31**, No. 4, 1045-62.

[54] Harris, B. (1973). The asymptotic distribution of the order of elements in symmetric semigroups, *J. Comb. Theory, Ser. A*, **15**, No. 1, 66-74.

[55] Harris, B. and Schoenfeld, L. (1967). The number of idempotent elements in symmetric semigroups, *J. Comb. Theory*, **3**, No. 2, 122–35.

[56] Ivchenko, G.I. (1973). The strength of connectivity of a random graph, *Theory Probab. Appl.*, **18**, No. 2, 396–404.

[57] Ivchenko, G.I. and Medvedev, Yu.I. (1965). Asymptotic representations of finite differences of a power function at an arbitrary point, *Theory Probab. Appl.*, **10**, No. 1, 139–44.

[58] Katz, M. (1955). Probability of indecomposability of a random mapping function, *Ann. Math. Statist.*, **26**, No. 3, 512–17.

[59] Katz, M. (1970). On the extreme points of a certain convex polytope, *J. Comb. Theory*, **8**, No. 4, 417–23.

[60] Knuth, D. (1969). *The Art of Computer Programming*, Vol. 2: Seminumerical Algorithms, Addison–Wesley, Reading, MA.

[61] Knuth, D. (1973). *The Art of Computer Programming*, Vol. 1: Fundamental Algorithms, 2nd edn, Addison–Wesley, Reading, MA.

[62] Knuth, D. (1973). *The Art of Computer Programming*, Vol. 3: Sorting and Searching, Addison–Wesley, Reading, MA.

[63] Kolchin, V.F. (1984). *Systems of Random Equations*, MIEM, Moscow (in Russian).

[64] Kolchin, V.F. (1986). *Random Mappings*, Optimization Software, New York.

[65] Kolchin, V.F. and Chistyakov, V.P. (1974). Combinatorial problems of probability theory, *Itogi Nauki i Tekhniki, Ser. Probab. Theory, Math. Statist., Theor. Cybern.*, **11**, 5–46 (in Russian).

[66] Kolchin, V.F., Sevast'yanov, B.A. and Chistyakov, V.P. (1978). *Random Allocations*, Wiley, New York.

[67] Kovalenko, I.N. and Levitskaya, A.A. (1986). *Selected Problems of Probabilistic Combinatorics*, Naukova Dumka, Kiev (in Russian).

[68] Kruskal, J.B. (1954). The expected number of components under random mapping function, *Amer. Math. Monthly*, **61**, No. 6, 392–7.

[69] van Lint, J.H. (1981). Notes on Egoritsjev's proof of the van der Waerden conjecture, *Linear Algebra Appl.*, **39**, pp. 1–8.

[70] London, D. (1971). Some notes on the van der Waerden conjecture, *Linear Algebra Appl.*, **4**, 155–60.

[71] Mahmoud, H.M. (1952). *Evolution of Random Search Trees*, Wiley, New York.

[72] Marcus, M. and Minc, H. (1964). *A Survey of Matrix Theory and Matrix Inequalities*, Allyn and Bacon, Boston.

[73] Meir, A. and Moon, J.W. (1970). The distance between points in random trees, *J. Comb. Theory*, **8**, No. 1, 99–103.

[74] Meir, A. and Moon, J.W. (1970). Cutting down random trees, *J. Austral. Math. Soc.*, **11**, No. 3, 313–24.

[75] Minc, H. (1978). Permanents, In: *Encyclopedia of Mathematics and Its Applications*, Vol. 6.

[76] Moon, J.W. (1969). Connected graphs with unlabeled end-points, *J. Comb. Theory,* **6**, No. 1, 65–6.

[77] Moon, J.W. (1970). Counting labeled trees, In: *Canadian Mathematical Congress,* Montreal.

[78] Moon, J.W. (1971). A problem of random trees, *J. Comb. Theory,* **10**, No. 2, 201–5.

[79] Moser, L. and Wyman, M. (1955). An asymptotics for the Bell numbers, *Trans. Roy. Soc. Canada,* **49**, Sec. 3, 49–54.

[80] Moser, L. and Wyman, M. (1955). On the solutions of $x^d = 1$ in symmetric groups, *Canad. J. Math.,* **7**, No. 2, 159–68.

[81] Moser, L. and Wyman, M. (1956). Asymptotic expansions, I, *Canad. J. Math.,* **8**, No. 2, 225–33.

[82] Moser, L. and Wyman, M. (1957). Asymptotic expansions, II, *Canad. J. Math.,* **9**, No. 2, 194–209.

[83] O'Neil, P. (1969). Asymptotic and random matrices with row-sum and column-sum restrictions, *Bull. Amer. Math. Soc.,* **75**, No. 6, 1276–82.

[84] O'Neil, P. (1970). Asymptotics in random 0, 1-matrices, *Proc. Amer. Math. Soc.,* **25**, No. 2, 290–5.

[85] Otter, R. (1948). The number of trees, *Ann. Math.,* **49**, No. 3, 583–99.

[86] Pavlov, Yu.L. (1988). Distributions of the number of vertices in strata of a random forest, *Theory Probab. Appl.,* **33**, No. 1, 96–104.

[87] Prokhorov, Yu.V. and Rozanov, Yu.A. (1973). *The Theory of Probability,* Nauka, Moscow (in Russian).

[88] Proskurin, G.V. (1973). On the distribution of the number of vertices in strata of a random mapping, *Theory Probab. Appl.,* **18**, No. 4, 803–8.

[89] Prufer, H. (1918). Neuer Beweis eines Satzes über Permutationen, *Archiv Math. und Phys.,* **27**, 742–4.

[90] Rabin, H. and Sitgreaves, R. (1954). *Probability Distributions Related to Random Transformations of a Finite Set,* Stanford University, Technical Report, **19A**.

[91] Rényi, A. (1959). Some remarks on the theory of trees, *Magyar Tud. Acad. Mat. kutato int. Közl.,* **4**, No. 7, 73–83.

[92] Rényi, A. (1962). Three new proofs and a generalization of theorem of Irving Weiss, *Magyar Tud. Acad. Mat. kutato int. Közl.,* **7**, No. 1, 203–14.

[93] Rényi, A. and Szekeres, G. (1967). On the height of trees, *J. Austral. Math. Soc.,* **7**, No. 4, 497–507.

[94] Riordan, J. (1958). *An Introduction to Combinatorial Analysis,* Wiley, New York.

[95] Riordan, J. (1960). The enumeration of trees by height and diameter, *IBM J. Res. and Develop.,* **4**, No. 5, 473–8.

[96] Riordan, J. (1962). Enumeration of linear graphs for mappings of finite sets, *Ann. Math. Statist.,* **33**, No. 1, 178–85.

[97] Riordan, J. (1968). *Combinatorial Identities*, Wiley, New York.

[98] Riordan, J. and Stein, P. (1972). Arrangements on chessboards, *J. Comb. Theory*, **12**, No. 1, 72–80.

[99] Robinson, R.W. and Schwenk, A.I. (1975). The distribution of degrees in a large random tree, *Discr. Math.*, **12**, No. 4, 359–72.

[100] Sachkov, V.N. (1971). Asymptotic normality of the distribution of the number of cyclic elements of idempotents in the symmetric semigroup, *Trudy MIEM*, Moscow, **14**, 180–90 (in Russian).

[101] Sachkov, V.N. (1971). The distribution of the number of fixed points of elements of a symmetric semigroup under the condition $\sigma^{h+1} = \sigma^h$, and the number of trees of height not exceeding h, *Theory Probab. Appl.*, **16**, No. 4, 661–74.

[102] Sachkov, V.N. (1971). Distribution of the number of distinct elements of a symmetric basis in a random mA-sample, *Theory Probab. Appl.*, **16**, No. 3, 494–505.

[103] Sachkov, V.N. (1972). Mappings of a finite set with limitations on contours and height, *Theory Probab. Appl.*, **17**, No. 4, 640–57.

[104] Sachkov, V.N. (1973). Enumeration problems of combinatorial analysis, In: *Problems of Cybernetics* (Proceedings of the Seminar on Combinatorial Mathematics AN SSSR), Sov. Radio, Moscow, 146–64 (in Russian).

[105] Sachkov, V.N. (1973). Random mappings with bounded height, *Theory Probab. Appl.*, **18**, No. 1, 120–30.

[106] Sachkov, V.N. (1973). Random partitions of sets with labeled subsets, *Matem. Sbornik*, **92**, No. 3, 491–502 (in Russian).

[107] Sachkov, V.N. (1974). Random partitions of sets, *Theory Probab. Appl.*, **19**, No. 1, 184–93.

[108] Sachkov, V.N. (1975). On the extreme points of the space of symmetric stochastic matrices, *Matem. Sbornik*, **96**, No. 3, 447–57 (in Russian).

[109] Sachkov, V.N. (1975). The system of distinct representatives of random sets, *Matem. Sbornik*, **97**, No. 3, 395–402 (in Russian).

[110] Sachkov, V.N. (1975). Random partitions, In: *Problems of Cybernetics*, Vol. 16, Part I (Proceedings of the Second All-Union Seminar on Combinatorial Mathematics), Sov. Radio, Moscow, 88–98 (in Russian).

[111] Sachkov, V.N. (1995). *Combinatorial Methods of Discrete Mathematics*, Cambridge University Press.

[112] Shepp, L.A. and Lloyd, S.P. (1966). Ordered cycle lengths in random permutations, *Trans. Amer. Math. Soc.*, **121**, No. 2, 340–57.

[113] Stepanov, V.E. (1966). Limit distributions of certain characteristics of random mappings, *Theory Probab. Appl.*, **14**, No. 4, 612–26.

[114] Stepanov, V.E. (1973). Random graphs, In: *Problems of Cybernetics* (Proceedings of the Seminar on Combinatorial Mathematics AN SSSR), Sov. Radio, Moscow, 164–85 (in Russian).

[115] Takács, L. (1965). A moment problem, *J. Austral. Math. Soc.*, **5**, No. 4, 487–90.

[116] Takács, L. (1967). On the method of inclusion and exclusion, *J. Amer. Statist. Assoc.*, **62**, No. 317, 102–13.

[117] Tanny, S. (1973). A probabilistic interpretation of Eulerian numbers, *Duke Math. J.*, **40**, No. 4, 717–22.

[118] Tutte, W.T. (1947). The factorization of linear graphs, *J. London Math. Soc.*, **22**, Part 2, No. 86, 107–11.

[119] Vatutin, V.A. and Mikhailov, V.G. (1982). Limit theorems for the number of empty cells in an equiprobable scheme for group allocation of particles, *Theory Probab. Appl.*, **27**, 734–44.

[120] Vatutin, V.A. (1993). The distribution of the distance to the root of the minimal subtree containing all the vertices of a given height, *Theory Probab. Appl.*, **38**, No. 2, 273–8.

[121] Vatutin, V.A. (1993). Branching processes and random trees, In: *Frontiers in Pure and Applied Probability*, Vol. 1 (Proceedings of the Third Finnish–Soviet Symposium on Probability Theory and Mathematical Statistics), VSP BV/TVP Science Publishers, pp. 256–69.

[122] van der Waerden, B.L. (1926). Aufgabe 45, *Über Deutsch. Math. Verein.*, **35**, 117.

[123] Wilf, H. (1966). On the permanent of a doubly stochastic matrix, *Canad. J. Math.*, **18**, No. 4, 758–61.

[124] Wilf, H.S. (1983). Three problems in combinatorial asymptotics, *J. Comb. Theory, Ser. A*, **35**, 199–207.

[125] Wright, E.M. (1973). The probability of connectedness of a large unlabeled graph, *Bull. Amer. Math. Soc.*, **79**, No. 4, 767–9.

Index

Absolute moment, 8
A-forest, 226
Aleksandrov, A.D., 49
$A\Lambda$-partition, 133
asymptotically normal
    sequence, 22

Bender, E., 32
Bernoulli numbers, 29
Bernoulli trials, 11
Bessel function, 15
binomial distribution, 11
Bonferroni inequalities, 14
Boole inequality, 2

Cauchy integral formula, 10
central limit theorem, 21, 22
  for sums of independent
     random variables, 22
  for triangular arrays, 23
central moment, 8
characteristic function, 18
  of a lattice distribution, 20
  of a normal distribution, 20
  of a Poisson distribution, 19
  of a uniform distribution, 18, 19
Chebyshev inequality, 8
complement of an event, 2
complementary event, 2
complete set of events, 2
complexity of a matrix, 169
conditional probability, 4
configuration, 82
connected component, 186
continuous random variable, 5
continuous random vector, 5
continuity theorem, 21
  for generating functions, 18

Curtiss' theorem, 25
cutting number of a vertex, 183
cyclic elements, 175

decrement of a permutation, 152
degree of a vertex, 174
density
  of a distribution, 5, 6
  of a normal distribution, 5
discrete random variable, 3
discrete space, 2
disjoint permutations, 50
distance between vertices, 180
distribution, 4
  binomial, 11
  hypergeometric, 12
  lattice, 19
  limiting, 18
  nondegenerate normal, 6
  normal, 5
  Pascal, 11
  Poisson, 12
  standard normal, 5
distribution function, 4
distribution law, 3
dominant function of a
  polynomial, 38
double generating function, 32

Egorychev, G.P., 49
GH-equivalence, 82
Euler numbers, 30, 31
event, 2
  certain, 2
  complementary, 2
  elementary, 2
  impossible, 2
  opposite, 2